Decolonising Heritage in South Asia

This volume cross-examines the stability of heritage as a concept. It interrogates the past which materialises through multi-layered narratives on monuments and other objects that sustain cultural diversity. It seeks to understand how interpretations of "monuments" as "texts" are affected at the local level of experience, even as institutions such as UNESCO work to globalise and fix constructs of stable and universal heritage.

Shifting away from a largely Eurocentric concept associated with architecture and monumental archaeology, this book reassesses how local and regional heritage needs to be balanced with the global and transnational. It argues that material objects and monuments are not static embodiments of culture but are, rather, a medium through which identity, power and society are produced and reproduced. This is especially relevant in South and Southeast Asian contexts, where debates over heritage often have local, regional and national political implications and consequences.

Reevaluating how traditional valuation of monuments and cultural landscapes could help aid sustainability and long-term preservation of the heritage, this book will be useful for scholars and researchers and Southeast Asian history, heritage studies, archaeology, cultural studies, tourism studies and political history as well.

Himanshu Prabha Ray is Honorary Professor of the Distant Worlds Programme, Ludwig Maximilian University in Munich, Germany. She is former Chairperson of the National Monuments Authority, Ministry of Culture in New Delhi, India, and former Professor in the Centre for Historical Studies, Jawaharlal Nehru University in New Delhi, India. Her research interests include Maritime History and Archaeology of the Indian Ocean, the History of Archaeology in South and Southeast Asia and the Archaeology of Religion in Asia. Her recent books include *Archaeology and Buddhism in South Asia* (2018), *Buddhism and Gandhara: An Archaeology of Museum Collections* (ed. 2018), *The Archaeology of Sacred Spaces: The Temple in Western India, 2nd Century BCE to 8th Century CE* (with Susan Verma Mishra, 2017), *The Return of the Buddha: Ancient Symbols for a New Nation* (2014) and *The Archaeology of Seafaring in Ancient South Asia* (2003). Her edited volumes include *Bridging the Gulf: Maritime Cultural Heritage of the Western Indian Ocean* (2016), *Indian World Heritage Sites in Context* (2014) and *The Sea, Identity and History: From the Bay of Bengal to the South China Sea* (with Satish Chandra, 2013).

Decolonising Heritage in South Asia

The Global, the National
and the Transnational

Edited by Himanshu Prabha Ray

Routledge
Taylor & Francis Group

LONDON AND NEW YORK

First published 2019 by Routledge

2 Park Square, Milton Park, Abingdon, Oxon, OX14 4RN

605 Third Avenue, New York, NY 10017

Routledge is an imprint of the Taylor & Francis Group, an informa business

First issued in paperback 2020

British Library Cataloguing-in-Publication Data
A catalogue record for this book is available from the British Library

Library of Congress Cataloging-in-Publication Data
A catalog record for this book has been requested

ISBN: 978-1-138-50559-9 (hbk)
ISBN: 978-0-367-73351-3 (pbk)

Typeset in Sabon
by Apex CoVantage, LLC

Contents

PART II
Case studies of World Heritage Sites in India

Figures

Contributors

I. **Wayan Ardika** is Professor in Archaeology, Faculty of Letters, Udayana University in Indonesia. He completed his undergraduate degree in Archaeology at Udayana University in Indonesia and his Master of Arts and Doctorate in Prehistory at Australian National University. He has been the Dean of Faculty of Letters, Udayana University (2003–2011), and Dean of Faculty of Tourism, Udayana University (1999–2002). Dr. Ardika's publications include *Burials, Texts and Rituals: An Ethnoarchaeological Investigation in North Bali, Indonesia* (ed., 2008); 'Blanjong: An Ancient Port Site in Southern Bali, Indonesia', in E. Hermann, K. Klenke and M. Dickhardt (eds.), *Form, Macht, Differenze. Motive und Felder ethnologischen Forschens* (2009); *Recent Studies in Indonesian Archaeology* (ed., 2010); and 'Sembiran: An Early Harbour in Bali', in J.N. Miksic, and G.G. Yian (eds.), *Ancient Harbour in Southeast Asia: The Archaeology of Early Harbours and Evidence of Interregional Trade* (2013).

Pia Brancaccio is Associate Professor and Co-director of the Art History Program at Drexel University in Philadelphia, Pennsylvania, USA. She earned her PhD in Indian Art History and Archaeology at the Universita' degli Studi di Napoli 'L'Orientale' in Naples, Italy. Prior to joining Drexel she held research positions at the Metropolitan Museum of Art in New York, Getty Research Institute in Los Angeles and at the Philadelphia Museum of Art. She has done extensive work on Buddhist art in ancient South Asia with a special focus on the Deccan plateau and the northwest of the Indian subcontinent. Dr. Brancaccio's publications include a monograph entitled *The Buddhist Caves at Aurangabad: Transformations in Art and Religion* (2010), two edited volumes entitled *Living Rock: Buddhist, Hindu and Jain Cave Temples in Western Deccan* (2013) and *Gandharan Buddhism: Archaeology, Art and Text* (with Kurt

Behrendt, 2006) as well as several articles in conference proceedings and academic journals (*Ars Orientalis, Archives of Asian Art, East and West,* and *South Asian Studies*).

Max Deeg studied Germanic Philology, Linguistics, Indology, Japanese Studies and Religious Studies and received his PhD and his Habilitation (a German professorial degree) at the University of Würzburg in Germany. He has taught in Germany, Japan, Austria and the UK, and is Professor in Buddhist Studies at Cardiff University in Wales, UK. His main research interests are the spread of Buddhism from South Asia to Central and East Asia and the reception history of Buddhist Studies. His most recent publications include a monograph on the early Buddhist history of Nepal and Buddhist foundation legends, and a co-edited book on pilgrimage and articles on the history of Buddhism.

Salila Kulshreshtha is the author of *From Temple to Museum in the middle Ganga Valley: Colonial Collections and Umā Maheśvara icons* (2018). She has worked on issues of urban heritage and heritage education with the Indian National Trust for Art and Cultural Heritage (INTACH) (2004) and with the Dr Bhau Daji Lad Mumbai City Museum in Mumbai, India (2011–2012). She has taught Art History, History and Humanities in Mumbai at Rizvi College of Architecture and Indian Education Society's College of Architecture (2012–2013) and in the USA at Old Dominion University and Virginia Wesleyan College (2005–2007). She is currently based in Dubai. Her research interests include the afterlives of sacred icons and shrines, issues of colonial archaeological interventions and the making of monuments and museums.

Swapna Liddle wrote her PhD thesis on cultural and intellectual change in 19th-century Delhi. She has been working for many years to raise awareness of the need to preserve historic buildings and neighbourhoods. Dr. Liddle is the author of *Delhi: 14 Historic Walks* (2011) and *Chandni Chowk: The Mughal City of Old Delhi* (2017). In addition, she is the convenor of the Delhi chapter of INTACH.

Lynn Meskell is Professor of Anthropology at Stanford Archaeology Center, Stanford University in Palo Alto, California, USA. She received her PhD from the University of Cambridge, UK, in 1997. Her current research and teaching interests include a broad range of fields, including ethnography in South Africa, Egyptian archaeology, identity and socio-politics, gender and feminism, and heritage ethics. As founding editor of the *Journal of Social Archaeology*, she has attempted to forge a vehicle for interdisciplinary dialogue, bringing together a wide range of scholars from diverse fields to constitute the editorial panel. Her new research focuses on the role of UNESCO in terms of heritage rights, sovereignty and international politics. She has recently published *A Future in Ruins: UNESCO, World Heritage and the Dream of Peace* (2018).

Indira Rajaraman holds a PhD in Economics from Cornell University in Ithaca, New York, USA. She has taught at University of Delhi in India; the Indian Institute of Management in Bengaluru, India; the University of Illinois Urbana-Champaign, USA; and the Indian Statistical Institute in Delhi, India. Her research papers span a wide range of development economics, including issues in fiscal policy, formal and informal financial institutions, and exchange rates and trade. She has been a member of the Thirteenth Finance Commission and of several committees that have shaped the process of financial and fiscal reform over the last two decades. She has been a regular columnist in the Indian financial press since 2001.

Himanshu Prabha Ray is Honorary Professor, Ludwig Maximilian University, Munich and Board Member, Oxford Centre for Hindu Studies, Oxford. She was former Professor, Centre for Historical Studies, Jawaharlal Nehru University, New Delhi where she taught until 2012. Subsequently she was the first Chairperson of the National Monuments Authority (Ministry of Culture, Government of India) until August 2015. Her research interests include Maritime History and Archaeology of the Indian Ocean, the History of Archaeology in South and Southeast Asia and the Archaeology of Religion in Asia. She is the Editor of the *Routledge Series on Archaeology and Religion in South Asia*.

Uthara Suvrathan is Adjunct Assistant Professor, Queens College, City University of New York, USA, after completing her PhD at the Department of Anthropology, University of Michigan in Ann Arbor, USA. Her dissertation was titled *Complexity on the Periphery: A Study of Regional Organisation at Banavasi, c. 1st – 18th century AD*. She has worked with archaeological and ethnographic collections in the Museum of Anthropology in Ann Arbor, Michigan, USA, assisting in the cataloguing and curation of artifacts, as well as with museum events, and supervising and training undergraduate student volunteers.

Preface

The five-year (2013–2018) Anneliese Maier research award of the Alexander von Humboldt Foundation hosted by the Distant Worlds Programme, Ludwig Maximilian University, Munich has been a tremendous source of strength that has allowed me to explore a range of themes through workshops and conferences. I am grateful to Professor Monika Zin for nominating me for the award and to Professor Jens-Uwe Hartmann for agreeing to collaborate with me on the theme of the research award: 'Cross-Cultural Dialogue: India and the Wider World in Ancient History'.

The first theme that I considered, Rethinking Gandhara, was explored through two international conferences: one held in Munich in March 2014 and the other in Delhi in April 2015. These meetings resulted in an edited volume titled *Buddhism and Gandhara: An Archaeology of Museum Collections* (2018). The theme of Heritage in South Asia was taken up at a workshop held in Delhi in August 2016, in collaboration with Nalanda International University, Rajgir, and the India International Centre, New Delhi. In July 2016, the archaeological site of Nalanda Mahavihara (Nalanda University) at Nalanda, Bihar was inscribed on the World Heritage List at the 40th session of UNESCO's World Heritage Committee held in Istanbul, Turkey from 10 to 20 July 2016. The partnership with Nalanda University, representing a revival of the "ancient Buddhist university", was both timely and appropriate for discussing issues relating to various aspects of World Heritage.

The present volume draws on papers from a three-day workshop held at India International Centre from 20 to 22 August 2016, which saw participation from members of the following institutions: School of Historical Studies, Nalanda University; Centre for Historical Studies, Jawaharlal Nehru University (JNU); INTACH, Delhi Chapter; and Aga Khan Trust for Culture. In addition, several independent researchers and scholars from universities around the world also participated, such as Professor Lynn Meskell, Stanford University; Professor Max Deeg, University of Erfurt; Dr. Pia Brancaccio, Drexel University; Dr. Uthara Suvrathan, Cornell University; Dr. Umakant Mishra, Ravenshaw University; and Dr. Deeksha Bharadwaj, Delhi University. I would like to express my thanks to Professors Gopa Sabharwal and Aditya Malik and their colleagues at Nalanda University – Murari Jha, Sraman Mukherjee, Samuel Wright, Abhishek Singh Amar, Andrea Acri, Noemie Verdon and Ranu Roychoudhuri – for their collaboration and participation, as also to Professor Vijaya Ramaswamy and Dr. Jyoti Atwal at JNU for their support. Ratish Nanda readily accepted the invitation to present the work being done by him and his team at Humayun's Tomb. Deeksha Bharadwaj and Suchi Dayal helped in wrapping up the proceedings by presenting eloquent summaries of the papers with wit and humour in the concluding session. Premola Ghose, Indira Rajaraman and Madhu Bhalla have been a constant source of encouragement. I am thankful to Shashank Sinha for his insights and for agreeing to publish the book. It is hoped that this would lead to greater involvement of the academic community in the several dimensions of the making and preservation of heritage.

1   Introduction

Himanshu Prabha Ray

The decolonisation of the mind is among the greatest challenges today's Indians have to face.
—Shashi Tharoor, Member of Indian Parliament[1]

Shashi Tharoor made this statement in an appeal to the Government of India to convert the Victoria Memorial Museum in Kolkata to a museum that depicts the atrocities committed by the British Raj against Indians and its impoverishment of the country. Regarding the latter, Tharoor made special reference to the construction of the railways in the subcontinent solely for the economic benefit of Britain. In this context one also needs to bring into discussion the destruction of archaeological sites which were looted for bricks and stone to be used in the laying of railway tracks: for example, in 1856 the third-millennium BCE site of Harappa was stripped bare of hard-baked mud brick to build 160 kilometres of the railway track between Lahore and Multan. Harappa was not the only site to suffer damage at the hands of railway contractors. The ancient mound of Rajghat near Varanasi was ravaged for bricks during extension and remodelling of the Kashi railway station. This wide-scale destruction was further compounded by the creation of an archaeological landscape in the subcontinent through intellectual frameworks that were derived from Victorian practices and provided teeth through legislation.[2] It is this decolonisation of the ancient heritage of South Asia that forms the central thesis of this book. The process of decolonisation must start with an unravelling of the categories created and publicised by British architects, such as James Fergusson (1808–1886), which continue to form the basis of art historical studies in the subcontinent.

Fergusson came to India to work for the family firm of Fairlie, Ferguson & Co. of Calcutta. Soon his interest shifted from merchandising to architecture, and from 1836 to 1841 he travelled to various parts of India, studying and documenting Indian architecture. After returning to London in 1845, his sketches were lithographed and published in a book entitled *Illustrations of the Rock Cut Temples of India*, which consisted of 18 plates.[3] In Fergusson's frame of reference, Indian architecture provided a missing link in the development of architecture in the world, especially the 12th–13th century flowering of architecture in Europe. In addition, even though India could never reach 'the intellectual supremacy of Greece, or the moral greatness of Rome' architecture in India was a living art which could inform study of developments in Europe in a variety of ways.[4] The emphasis on architecture was necessary, since there was a lack of historical texts in India and post-5th-century Indian history could only be studied through monuments and inscriptions. Fergusson's classification was produced within a racial-religious framework; for example, the limits of Dravidian architecture were defined by the spread of people speaking Tamil, Telugu, Malayalam and Kanarese.[5]

Fergusson's work established a link between architectural form, ethnicity and religious affiliation. Within this framework, an apsidal shrine such as that of the Durga temple at Aihole, northern Karnataka, could be associated with Buddhism only. This 8th-century Durga temple was dedicated to Surya, which Fergusson explained by underscoring its later conversion to Hinduism. By the 1860s the Durga temple was featured 'as an inglorious, structural version of a Buddhist *chaitya* hall, appropriated by Brahmanical Hindus and buried under rubble at a site of the ancient Chalukya dynasty'.[6] As a result of subsequent investigation by Gary Tartakov that considers not only the plan of the temple, but also its rich imagery, it is now evident that the Durga temple is one of nearly 150 temples built across 450 kilometres of the Deccan in the 7–8th centuries CE. This temple is the largest and most lavishly constructed monument dating to around 725–730 CE (Figure 1.1).

A linear view of religious development emerged in the 19th century that attributes the beginnings of stone architecture dated to the 4th–3rd centuries BCE largely to Greek influence. The connection between architectural form and religious change was firmly established and the quest for chronology securely rooted architecture within linear time. The linear development of Buddhist–Jain–Hindu architecture accepted notions of origins and decline; it assumed antagonism between the different religions of the subcontinent, and ruled out coexistence.[7] More importantly, religious shrines were converted into monuments that proclaimed the country's ancient past – a past that was to be

Figure 1.1 Durga temple at Aihole, based on an apsidal plan.
Source: Courtesy of the author.

sanitised by keeping all human habitation at bay, and one which was to be landscaped and beautified by the creation of gardens and lawns. These monuments have continued as a legacy of the British Raj. The Archaeological Survey of India was established in 1861 and charged with safeguarding this heritage. In addition, laws were created for its preservation and protection, such as the Ancient Monuments Preservation Act of 1904, which was re-validated post-Independence in 1958 and more recently in 2010.[8]

The report of the Comptroller and Auditor General of India for the year ended March 2012, which contains the results of a performance audit conducted on several institutions under the Ministry of Culture, states that as per budget estimates for the year 2011–2012, approximately 33% of the total budget of the Ministry of Culture was utilised for the functioning of the ASI. Another 6% was given to the seven major museums of the country.[9] Together, these institutions are the repository of the country's heritage and treasures, yet there has been little academic scrutiny of their vision and its implementation.

This chapter starts with a discussion of the "discovery" of India's heritage and highlights the inadequacies in inscribing the 27 cultural sites on the World Heritage List as "monuments" with two examples, viz. Sanchi and Rani-ki-Vav, which have focussed narrowly on art and architecture thereby neglecting the dynamic history of the sites. This first section is followed by a discussion on historiography in the second, while in the third and fourth sections the focus is on the use of heritage in fostering cross-cultural connections, such as transregional heritage of South Asia, which needs to be promoted on the global stage to build sustainable ties with countries across political boundaries. The final section provides an outline of the book.

World Heritage Sites in India: from monuments to cultural landscapes

The central argument of this book is that material objects and monuments are not static embodiments of culture but are, rather, a medium through which identity, power and society are produced and reproduced. The processes through which transformations take place are vital to an understanding of the role that heritage and the past could play in South Asia, especially in preserving cultural plurality and building sustainable networks. India has played an active role in the World Heritage Committee since 1977, when the country ratified UNESCO's 1972 Convention for the protection and preservation of the world's natural and cultural heritage considered to be of outstanding value to humanity. India has so far inscribed 27 cultural, seven natural and one mixed site on the World Heritage List. The latest addition to the list (on 8 July 2017) is the historic city of Ahmedabad, the first World Heritage city in India.

Twelve World Heritage Sites in India date to the ancient period, ten to the mediaeval and only four to the colonial period. Bhimbetka in central India is the only pre-historic site. Bhimbetka is also one of the first sites inscribed as a Cultural Landscape in 2003, while two others, such as the Apatani cultural landscape and the Cold Desert cultural landscape of Ladakh are on the Tentative List. An emphasis on the ancient past is evident, but perhaps what is required is a policy for selection of sites for UNESCO inscription[10] which would involve dialogue and discussion with several stakeholders, including those living close to the sites. As it stands, the decision to propose sites for inscription to the World Heritage List lies with officials of the Ministry of Culture and nodal agencies such as the Archaeological Survey of India. It is suggested here that this lack of public dialogue has resulted in a continuation of the hierarchy of monuments that was prescribed in the 1904 Act.

The most famous monuments of ancient India are the free-standing pillars of the 3rd century BCE Mauryan emperor, Ashoka, which bear his inscriptions and are crowned by stupendous pillar capitals, such as the lion capital at Sarnath, which is independent India's national emblem. The aesthetics of the sandstone pillars and their polish brought them to the notice of early visitors and travellers to the sub-continent, who not only described them in glowing terms, but also sketched and painted them before photography became the norm. The first pillar of Ashoka was found in the 16th century by Thomas Coryat (1577–1617) in the ruins of ancient Delhi at Ferozeshah Kotla. This ancient pillar had been shifted in 1356 by a mediaeval ruler, Feroze Shah Tughluq (1351–1388) to adorn his palace (Figure 1.2). Initially Coryat assumed from its polish that the pillar was made of brass, but on closer examination he realised it was highly polished sandstone with upright script that he thought resembled a form of Greek (Figure 1.3).[11] He credited Alexander the Great with setting up the pillar to commemorate his victory over Porus.[12] The reference to the Greeks is significant, since it was the search for cities established by Alexander

Figure 1.2 Ashokan pillar at Ferozeshah Kotla, Delhi.

Source: Courtesy of the author

Figure 1.3 Close up of the Ashokan pillar at Ferozeshah Kotla.

Source: Courtesy of the author

and the military campaigns conducted by him that provided an impetus for the beginnings of archaeology in colonial India, especially in the northwestern part of the subcontinent.[13]

The idea of government-sponsored archaeology was largely the result of Alexander Cunningham's (1814–1893) bold initiatives and resulted in the setting up of the Archaeological Survey of India in 1861. In his search for sites associated with the life of the historical Buddha, Cunningham relied on accounts of the Chinese pilgrims Faxian and Xuanzang who travelled to India in the 5th and 7th centuries respectively. While he resorted to Sri Lankan chronicles such as the 4th-century *Mahavamsa* to identify the role of the Mauryan emperor, Ashoka, in propagating Buddhism, he had little interest in comprehending the survival of narratives relating to Ashoka in later Buddhist texts and inscriptions – an issue stressed by later researchers. For example, we know of several pillars and rocks which bear inscriptions of later rulers, such as the Junagadh rock edict in Gujarat, which also has inscriptions of the Kshatrapa ruler Rudradaman and 4th-century King Samudragupta of the Gupta dynasty. It is also evident from later sources that many of the monastic establishments that grew up around the pillar edicts (at Sarnath, for example) preserved the association of the Buddhist emperor Ashoka with the pillars. An 11th-century pedestal inscription found at Sarnath records the restoration of the stupa of Ashoka at Sarnath by two brothers from Gauda who were identified with eastern India.[14]

In addition to the *Mahavamsa* used by Cunningham, the narrative on Ashoka occurs in several Buddhist texts, starting as early as the *Ashokavadana*, which was written in Sanskrit in the 2nd century CE, 500 years after Mauryan rule. The text formed a part of the *avadana* genre and was perhaps compiled in northwestern India. Ashoka is eulogised in Buddhist writing for visiting places associated with the life of the Buddha and sign-posting them either with his pillars or by establishing stupas. One of the concerns of the *Ashokavadana* was to define the nature of Buddhist kingship and the extent to which its generosity impacted the monastic community. The *Ashokavadana* is important in that it forms the basic version of the Ashoka legend as it circulated in northwest India and found its way into Central Asia, China, Korea, Japan and Tibet. It also inspired several later writings, including the 16th-century work titled *History of Dharma* of the Tibetan monk Taranatha.[15] 'In Theravada countries such as Sri Lanka, Thailand, Laos and Burma, he [Ashoka] was and still is portrayed as a paradigmatic ruler, a model to be proudly recalled and emulated'.[16] It is significant that the Pha That Luang, or the Great Stupa, in Vientiane, the capital of the Lao People's Democratic Republic, is a national symbol and also the most sacred monument in the country. Popular belief associates the site with the location of an ancient stone pillar said to have been erected by Emperor Ashoka and containing the relics of the Buddha. The present stupa was built by King Setthathirat in 1566 as he shifted his new capital from the town of Luang Prabang and connected it to the narrative of the origins of Buddhism in the country.[17] The adoption of Ashoka as a model emperor to be followed is a phenomenon evident in several parts of Southeast Asia.

In India none of the pillars, capitals or edicts has ever been nominated for World Heritage status. The closest that one can get to preserving the association with the Mauryan emperor on a global platform is at Sanchi, 40 kilometres from Bhopal, the capital of Madhya Pradesh. Sanchi is the site of an Ashokan pillar and extensive remains of Buddhist monasteries and temples that date from the 3rd century BCE to the 12th century CE (Figure 1.4). Though it is not associated with events in the life of the Buddha, Sanchi is well-known for a Schism edict of Ashoka. In his Schism edicts at Sanchi and Sarnath, Ashoka cautions monks against creating dissensions within the *sangha* (community). Clearly Sanchi had a large and powerful monastic settlement, which is evident from inscriptions on reliquaries.

Figure 1.4 Sanchi Stupa 1 near Bhopal in central India.

Source: Courtesy of the author

After Cunningham completed excavations at Sarnath and other sites in north India, he turned his attention to Sanchi (the Bhilsa topes), which is situated on a low ridge on the banks of the Betwa river. He excavated this site in January and February 1851 along with F.C. Maisey. Cunningham's primary interest was in relics and relic caskets and at both Sarnath and Sanchi he dug shafts through the centre of the stupas. In contrast to Sarnath, Cunningham was more successful in collecting a large hoard from the several stupas that he opened at Sanchi and its vicinity. He found relics of the Buddha's disciples Sariputta and Moggalana from Stupa 2 at Satdhara and Stupa 3 at Sanchi. Cunningham discarded the stone boxes in which the caskets had been placed and transported the caskets containing relics to London. The stone boxes were subsequently located during excavations conducted by John Marshall in the early 20th century, but the relics, along with the caskets, seem to have been lost.[18] Marshall theorised that the relics of the two monks had been shifted to Sanchi when additions were made to the monastic structures.[19] The number of inscribed relic caskets (which may be identified with Buddhist monks and teachers) found at Sanchi is striking.[20] Not all stupas, however, contained relics; Cunningham lists several that he opened without any results.[21] He explained the importance of his discoveries as follows:

> As the opening of the Bhilsa Topes has produced such valuable results, it is much to be hoped that the Court of Directors will, with their usual liberality, authorise the employment of a competent officer to open the numerous Topes which still exist in North and South Bihar, and to draw up a report on all the Buddhist remains of Kapila and Kusinagara, of Vaisali and Rajagriha, which were the principal scenes of Sakya's labours. A work of this kind would be of more real value for the ancient history of India (the territory of the Great Company) than the most critical and elaborate edition of the eighteen *Purāṇas*.[22]

In 1919, the Bhopal Durbar initiated negotiations for the return of the relics that Cunningham had taken from Sanchi – negotiations that would continue into post-Independence India. Eventually, in February 1947, the relics were presented to the Mahabodhi Society in London; however, they were not housed in their original caskets, but in wooden reproductions.[23] The Government of India strongly objected to the reproductions and insisted on the original caskets being sent.[24] The relics in their original caskets then travelled through several countries, such as Sri Lanka and Burma, and were brought to Calcutta from Colombo. They then were placed on a special train that toured several parts of the country. Finally, in November 1952, the relics were enshrined in a *vihara* (monastery) specially built for them at Sanchi. Thus, the history of the "archaeological discovery" of Sanchi is as fascinating as the history of its art and architecture, though it is the latter that has been used as criteria for inscribing the site in 1989 as one of the oldest Buddhist sanctuaries, thereby missing out on an opportunity to celebrate the later history of the site.

The Sanchi Survey Project, which comprised an extensive survey of the cultural landscape around the well-known Buddhist complex, mapped roughly 720 square kilometres of countryside. In the process it revealed 35 Buddhist sites, a network of dams and 17 irrigation works, and 145 habitation settlements.[25] Several stupa sites such as Sonari, Satdhara, Morel Khurd and Andher are situated within 15 kilometres from Sanchi. About 8 km north of Sanchi is the ancient city of Vidisha, which is located at the confluence of the Betwa and Bes rivers. Udaygiri Hill, which has preserved remains of Shaiva, Vaishnava and Jain sculptures, is 1.5 kilometres to the west of the ramparts of Vidisha. Rock shelters and pre-historic sites add to the diversity of the Sanchi area. Undoubtedly this cultural landscape is significant for an appreciation of the architectural history of the Buddhist monuments of Sanchi.

As with Sanchi, another case of marginalisation of the cultural landscape may be quoted from a more recent inscription in 2014 on the World Heritage List of the Queen's stepwell built by queen Udayamati, which is locally known as Rani-ki-Vav (a Hindi version of the Gujarati name Ranki Vav) on the World Heritage List in 2014 (Figures 1.5 and 1.6).

> Rani-ki-Vav, on the banks of the Saraswati River, was initially built as a memorial to a king in the 11th century CE. Stepwells are a distinctive form of subterranean water resource and storage systems on the Indian subcontinent, and have been constructed since the third millennium BCE.[26]

Gujarat has a long history of water architecture starting from the Harappan period when wells were built with steps for accessibility at several sites, including at Dholavira.[27] Located on the outskirts of the city of Patan in North Gujarat, about 2 kilometres to the northwest of the modern city, Ranki-vav (correct nomenclature in Gujarati) was built in about 1050 CE by Queen Udayamati in memory of her husband, the founder of the Solanki dynasty, and is the most elaborately constructed and richly embellished stepwell encountered

Figure 1.5 Inside view of Ranki Vav near Patan in Gujarat showing the different stories of the stepwell.

Figure 1.6 Varaha *avatara* panel at Ranki Vav.

Source: Courtesy of the author

to date.[28] Close to Ranki Vav is a large reservoir named the Sahas-tralinga Talav for the 1,000 Shivalingas or shrines it was supposed to have had on its banks. Today, not a single Shivalinga or shrine can be seen, though ruins of what could have been a temple exists at the end of one embankment with some pillars still standing upright. The Sahastralinga Talav covered an area of 5 square kilometres and was fed by waters from the Saraswati river. The reservoir's embankments were made of stone and had steps leading to the water. The tank also had a water filtering system with a cistern that was about 40 m in diameter. The volume of water entering the cistern could be controlled through inlet pipes. Legend associates the tank with the Odh tribe, a tribe of pond diggers, and the sacrifice of Jasma Odan, the wife of Rooda, who committed *sati* after her husband was killed by the king. Her curse made the tank waterless. Since the 19th century, the legend of Jasma Odan has been revived as folk theatre and continues to enthral audiences.

Thus, these reservoirs and stepwells are not merely water bodies or wells for facilitating access to water in a dry and arid zone such as Gujarat. Given the intricately sculpted nature of many of the stepwells and the large number of deities represented, these have often been considered as subterranean temples, though the unique feature of the Ranki Vav as a memorial built by the queen for her deceased husband has yet to be explored. Another interesting facet is the fact that several stepwells were built by women of means, such as queens, wives of rich merchants, ordinary women, courtesans, etc. There also seems to be a close association between the waters and the cult of the mother goddess. In several cases, there are shrines to the goddess either within the precincts of stepwells or close to them. Stepwells also became a favourite subject in folklore and have formed the setting for tales of love and betrayal, courage and sacrifice.[29]

Until 1950 the Ranki Vav stepwell was practically unknown as it lay covered with silt, debris and earth. Columns on the first tier had been hauled off to build the nearby 18th-century Bahadur Singh *ki vav*, now completely encroached. It was in the middle of 1950 that archaeological excavations were initiated by the Archaeological Survey of India and the stupendously sculpted stepwell was unearthed after painstaking work. The well extends over 4.68 hectares but is believed to be part of a larger complex comprising bath-like structures and temples surrounded by flowing water. It was generally accepted that archaeological excavations in the area would yield more structures. In November 1951, the Ranki Vav was notified as a nationally protected monument and since 2014 has World Heritage status.

There is no doubt about the magnificence of the stepwell. This chapter suggests that the appeal of the Ranki Vav to the local communities and visitors would be increased manifold if it was located and studied within its larger context of local history and narratives. It is important that the stepwell be viewed not merely as an isolated water body, but as an indicator of a larger worldview that needs to be preserved and protected as the country's unique heritage, an objective unfortunately missing from most of India's World Heritage nominations.

History and heritage: definitions and historiography

How is "heritage" to be defined and distinguished from "history", though contemporary print media often uses the terms interchangeably?[30] Heritage specialists link heritage with place and define it as a process used in the creation of identity in a variety of ways. Thus, heritage is seen as an actively constructed public discourse of the past, which is fluid and changeable.[31] Heritage is deeply enmeshed with materiality and the present. However, this tangible past cannot be separated from intangible beliefs and resonances.[32] The larger issue of relevance is: Whose interpretation of cultural heritage predominates, and how is this interpretation mobilised, especially with reference to monuments and archaeological sites proposed for nomination at the global level?

The distinction between history and heritage underscores the significance of "monuments" as "texts" to be read through multi-layered narratives that sustain cultural diversity. The chapters in the book examine the idea of heritage as it has evolved from a European-originated – and until recently largely Eurocentric – concept associated with architecture and monumental archaeology to integrate diverse cultural forms and values in the historical and socio-cultural contexts of non-European societies. Monuments in South Asia have generally been studied in terms of architecture and imagery, or regarding chronology and patronage, and more recently within debates of generation of colonial knowledge, but seldom regarding cultural plurality and memory of the community. Two aspects of monuments in Asia are striking: their diversity and their interconnectedness. However, these have seldom found space in academic studies or in World Heritage nominations.

Though the theme of heritage and monuments in the colonial period has been addressed in several studies,[33] it continues to be a somewhat under-researched topic for the post-Independence period. Bridget Allchin, F.R. Allchin and B.K. Thapar edited a book that draws on papers presented at a seminar held in Cambridge in the run up to the first Festival of India in London in 1985 and covers both natural and cultural conservation.[34] Allchin's chapter in the volume refers to loss of cultural heritage in both rural and urban areas resulting from expansion of agricultural operations, infrastructure development, and quarrying for stone and brick.[35] In addition, he writes, 'The uncontrolled growth of tourism, both domestic and foreign, and of its infrastructure, must also be regarded as a potential threat to the cultural heritage'.[36]

One of the fundamental assumptions of Tapati Guha-Thakurta's analysis of India's religious structures and historical monuments and their contentious histories of identity politics in colonial, nationalist and post-colonial contexts is the power of knowledge in the colonial, and later national, as well as the institutional structures of its production. Two of the case studies in the book that have generated a growing body of secondary literature focus on Bodhgaya and Ayodhya.[37] An issue missing from the discussion is the cultural-historical framework within which colonial knowledge was constructed and embedded through legislation, such as the Ancient Monuments Protection Act passed in 1904 when John Marshall was the Director General of the Archaeological Survey of India. A rich corpus of secondary literature is available for the site of Bodhgaya that highlights the political agendas of persons involved in disputes around the site and the British handling of it.[38] An assumption of Guha-Thakurta that the conservation of monuments 'amounted to their effective museumisation'[39] has been countered by Deborah Sutton, who shows that the colonial state used the authority provided by the 1904 Act 'to impose conservation to a degree unknown in Europe, and especially in Great Britain, where the passage of similar acts was subject to a variety of caveats'.[40] This was not always successful and in some cases the interests of the deity of the religious shrine superseded other concerns, as a result of local resistance.

Santhi Kavuri-Bauer has written on the shifting spatiality of Mughal monumental architecture in northern India.[41] Asserting that 'monumental environments materialise power relations, influence the social ordering of a nation, produce us as subjects, and finally, and more positively, provide us with a critical space to create, resist and endure in our everyday lives', Kavuri-Bauer deploys evidence from the literary to the psychoanalytical to demonstrate the contributions of Mughal architecture to the developing political and social orders in the 19th and early 20th century.[42] She argues that these Mughal sites, built between the 16th and 17th centuries, have at different junctures in modern Indian history repeatedly been produced as social spaces in which Indian nationhood was enacted or its very possibility contested. The book covers the period from the 18th century to the post-Independence period, with the last chapter (i.e. Chapter V) devoted to monuments as heritage and as centres of tourism under the Second Five-Year Plan (1956–1961) when special funds were allocated for infrastructure development around monuments.

Certain contradictions in the text need to be ironed out, for example the statement that post–Indian Independence in 1947, Mughal monuments were radically and singularly reframed as national heritage.[43] Mughal monuments were already on the list of nationally protected monuments at the time of Indian Independence in 1947. It has been suggested that in the colonial period Mughal buildings were specifically chosen for preservation on a priority basis, as the English saw themselves as their natural and legitimate successors.[44] To what extent can religious affiliation be attributed to monuments? What Kavuri-Bauer misses is the extent to which the 19th century created iconic "Hindu" and "Muslim" spaces to the exclusion of others, thereby exacerbating religious identities of structures.[45] For example, the Chaukhandi mound at Sarnath yielded a large number of Buddhist sculptures in 1905, though a 16th-century octagonal brick tower exists on top of it. Govardhan, son of Raja Todarmal, constructed the tower to commemorate Emperor Humayun's visit to the place. An inscription recording the event refers to it as a lofty tower reaching to the blue sky.[46] Besides, even by Kavuri-Bauer's own admission financial support for the development of tourist potential of nationally protected Mughal monuments came after the Second Five-Year Plan, several years after independence.

Perhaps it would be appropriate to conclude this section with a quote from a former Secretary, Ministry of Culture, Government of India:

> Where India's built heritage is concerned, the ASI has done its bit, but it is unfair to expect this terribly understaffed organisation to protect all the 3,650 notified monuments. More than half of these could either be de-notified with no great harm to heritage or handed over to the local community to involve them, instil pride and ensure protection. This would permit the ASI to concentrate on a limited number that would hopefully be better maintained, even within its meagre budget. A few years ago, we had pleaded before the Planning Commission for a raise in public funding, arguing that the central government spends just 13 paisa per 100 rupees of its budget on the culture sector, which is indeed very low.[47]

Thus the reality and financial support for culture and heritage forms no part of the national agenda, despite public speeches to the contrary, as shown in the next section.

Heritage as providing building blocks for the present

The importance accorded to monuments as building blocks of partnerships for the present and the future is evident from the fact that the first-ever Asian Relations Conference was held in Delhi from 23 March to 2 April 1947, prior to Indian independence on 15 August 1947. It was attended by approximately 250 delegates from some 32 countries, observers from across the globe and about 2,000 local visitors.[48] The conference venue was the 16th-century historic fort of Purana Qila, which was built by the second Mughal emperor, Humayun (1508–1556), but subsequently captured by the Afghan Sher Shah Suri (1486–1545) who ruled Delhi briefly in the mid 16th century, before ceding it back to the Mughals. The Purana Qila, however, is much more than merely a mediaeval structure.

Archaeological excavations conducted within its ramparts by the Archaeological Survey of India from the 1950s onwards confirm that it represents the site of Delhi's earliest settlement and dates to the first millennium BCE. It represents both continuity of habitation at the site and also the essence of the city. 'A Mughal chronicler from the time mentions that Humayun's fort was built on the hallowed site of Indraprastha, a story that sits in the shimmering realm between myth and history'[49] – or between historical memory as enshrined in the epics, in this case the Sanskrit epic, the *Mahabharat* and the constant restructuring of the site by subsequent rulers. Thus the Purana Qila is an appropriate example of multi-layered histories that characterise several locations and places in the subcontinent, histories which have been reduced to a single monolithic category, as origins and chronology took precedence over historical memory and reuse. This also raises the issue of political will and state policy towards cultural heritage as a tool of cultural diplomacy.[50] How effectively has state policy been implemented? Has it even found a public platform for discussion? This complete neglect of dialogue among stakeholders is evident from the next example.

The Indian History Congress, at its 75th annual session held at Jawaharlal Nehru University from 28 to 30 December 2014, passed a resolution related to the violations of rules of preservation and restoration in the work carried out by a private agency [Aga Khan Foundation] at Humayun's Tomb, Delhi. It urged an enquiry into how this happened and demanded that all preservation, restoration work by similar private agencies be stayed until the whole matter is clarified.[51] To what extent is this resolution based on an uncritical reading of John Marshall's 1923 *Conservation Manual,* which has served as the primary guide for the conservation of monuments both by the Archaeological Survey of India and State Departments of Archaeology and has only recently been revised and submitted for approval to the Ministry of Culture in 2014? It is suggested that the critique of work being done at Humayun's Tomb perhaps draws on the lure of the picturesque and the atmosphere of timeless decay which formed the basis of colonial aesthetics as represented in the aquatints and paintings of British artists.[52]

In this introductory chapter, the focus is on World Heritage Sites of South and Southeast Asia, so as to move beyond the categorisation of single or group of monuments located within national agendas to address the larger issue of connectivity and mobility between the regions. Farish Noor, for example, has argued that since the construction of the nation state was a colonial legacy and artificial to begin with, this can be reconstructed and deconstructed to form new linkages between the two regions of South and Southeast Asia, which reflect or are closer to its pre-colonial past when it was a contiguous borderless region.[53] Is this possible through the global network of UNESCO and the World Heritage Committee?

Moving beyond national boundaries

Strategies to protect and conserve cultural heritage internationally have been successfully developed over the last several decades through UNESCO's 1972 World Heritage Convention. India signed the Convention in 1977. The Archaeological Survey of India functions as a nodal agency for nomination of World Heritage Sites to UNESCO, though the list is ultimately cleared by the Ministry of Culture. Even though UNESCO has over the years widened the parameters of defining culture, and its Universal Declaration of Cultural Diversity (adopted in 2005) states that cultural diversity and plurality need to be protected and preserved, this has not filtered down to the ground level where "monuments" continue to be inscribed within the somewhat restrictive definitions of the 1904 Ancient Monuments Protection Act and its subsequent avatars. At the global level, UNESCO's several charters remain fragmented and diverse and lack a comprehensive enabling vision. The focus continues on the conservation of monuments, rather than on promoting a universal understanding of cultural diversity and plurality. In this, most countries continue to follow their colonial legacy. These issues will be investigated in this final section through an analysis of heritage and archaeological sites in South and Southeast Asia.

The World Heritage Centre (WHC) was established in 1992 to act as the secretariat and coordinator within UNESCO for all matters related to the World Heritage Convention. The WHC organises annual sessions of the World Heritage Committee and provides advice to States Parties in the preparation of site nominations. The WHC, along with the Advisory Bodies, also organises international assistance from the World Heritage Fund and coordinates both the reporting on the condition of sites and the emergency action undertaken when a site is threatened.[54] Securing a seat on the World Heritage Committee is seen by many countries as one way to effectively raise their profile. India has served thrice on the World Heritage Committee: from 1985 to 1991; from 2001 to 2007; and more recently from 2011 to 2015. India's contribution to the UNESCO's World Heritage Fund was to the tune of $17,435 US dollars in 2011–2012. However, on UNESCO's request, India also made an extra budgetary contribution of $100,000 US dollars to the World Heritage Committee for specific projects.[55] An issue to which no clear answers are available is the following: What was the state's policy during periods of India's membership in the WHC and to what extent did this membership result in successfully forging partnerships and alliances?

The earliest monuments to be inscribed on the World Heritage List in India in 1983 included the Agra fort, Ajanta caves, Ellora caves and the Taj Mahal, to be followed in 1984 by the group of monuments at Mahabalipuram and the Sun temple at Konark. In the first two years, India nominated sites only in the category of "culture"; a year later, in 1985, natural sites were included in the list. Of these, the site of Ellora, in the Aurangabad district of the present state of Maharashtra in India, is an appropriate example of cultural and religious diversity.[56]

The earliest cave excavation at Ellora began in the late 6th century CE and was dedicated to Shiva, followed by Buddhist and Jain caves over the next several centuries until the 10th century CE. Though a majority of Ellora's Hindu excavations are dedicated to Shiva, the two exceptions are Caves 14 and 25, which appear to have been temples to Durga and Vishnu (or possibly Surya), respectively. Cave 16, famous as the monolithic rock-cut Kailashanatha temple dedicated to Shiva, is admired for its conceptualisation and sculptural exuberance. The Shaiva caves share several architectural features with the 12 Buddhist caves at Ellora, which were excavated from 600 to 730

CE. They document the development of Vajrayana imagery from the simple delineation in Cave 6 to the elaborate forms of Cave 12. Much of the excavation activity for the Jain cave temples was conducted during the 9th and 10th centuries CE, a time when the Rashtrakutas had attained paramount sovereignty in the region. Although the Archaeological Survey of India has categorised the Jain monuments into five separate cave complexes (Caves 30–34), there are in actuality 23 individual cave temples, nearly all of them containing a shrine and rock-cut Jina image.[57]

Three kilometres from the caves at Ellora is Khuldabad, known as the valley of saints as it is said to contain the graves of 1,500 Sufi saints, as well as the tomb of the Mughal Emperor Aurangzeb in addition to those of his sons and his generals. Marking the Chisti establishment at the site are the tombs of Sayyad Burhan-al din, a Sufi saint who died in 1344, and the mausoleum of Sayyed Zain ud din, another saint highly revered by the Muslims. On the east side the latter contains a number of verses inscribed from the Quran and the date of the saint's death in 1370 CE. These tombs are important markers of the 14th-century Sufi tradition of Nizamuddin Auliya, which went from Delhi to the Deccan and established itself in Khuldabad.[58] Ellora is by no means the only example of religious pluralism in South Asia, but is instead one of the many sacred places that preserve diverse historical memories. However, the reasons and the criteria adopted for the inscription of the site were rather different – and cultural plurality was not one of them.

Cultural plurality as demonstrated by the monuments and their landscapes is important if these sites are to be placed in context, and any discussion of these as isolated structures detracts from their value. The theme of inter-member-state or transnational nominations highlights the possibilities that may exist in South Asia. A beginning in this regard has already been made with the Silk Roads World Heritage Serial and Transnational Nomination, which brought together several Central Asian Republics along with China to share knowledge and expertise and to highlight connections that existed in antiquity.

Increasingly, maritime historians agree that the history of the seas that unite South and Southeast Asia should be discussed as "connected history" across porous borders, linked through boat-building traditions, community networks and cultural practices. While the history of maritime trade, trading commodities and ports has received attention and there is a rich secondary literature available on these themes, this is not the case with subjects such as coastal architecture, heritage practices and boat-building traditions. It was with a view to building

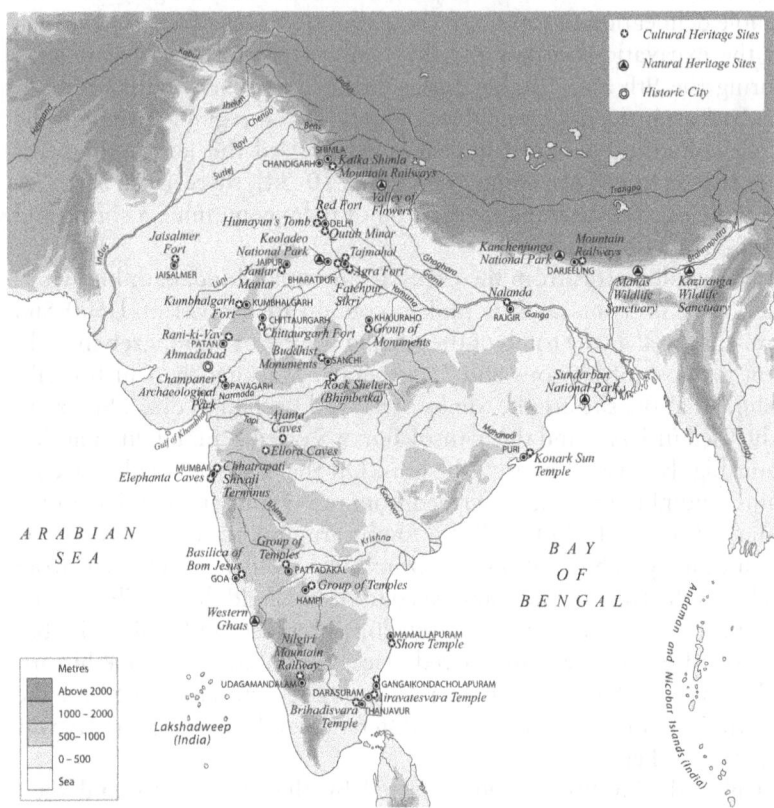

Figure 1.7 Map showing World Heritage Sites in India.

Source: Map drawn by Uma Bhattacharya

bridges across the seas that the Ministry of Culture, Government of India proposed the Project Mausam at the 38th World Heritage Session at Doha, Qatar on 20 June 2014.[59] It is unfortunate that there has been very little progress on the nomination over the last three years, in spite of the fact that rupees 15 crores was sanctioned by the Ministry of Culture in 2015 and two institutions under the ministry (i.e. the Archaeological Survey of India and the Indira Gandhi National Centre for the Arts) were entrusted with the task of coordinating Project Mausam with other countries and interested States Parties.[60]

It is significant that many of the World Heritage Sites in Southeast Asia are in coastal areas or river valleys that provide communication between the coast and the interior and are known to have participated in the maritime networks historically. It is time to underscore their connectedness within the larger oceanic system and to bring these sites into dialogue with each other across political boundaries (Figure 1.7). Thus, a transnational perspective is crucial both to UNESCO's World Heritage project as also for the protection and preservation of national heritage.

Outline of the book

The chapters are divided into three sections: Part I focuses on World Heritage Sites through several perspectives; Part II presents case studies of World Heritage Sites in India; and Part III, titled "Transnational Heritage", presents a discussion of heritage beyond national frontiers. The first chapter, by Lynn Meskell, argues that unresolved tensions surrounding the UNESCO's foundation in 1945 remain the same as those confronting global governance of the world's natural and cultural heritage in 2016: cultural and religious differences, economic inequality, and fundamentally divergent understandings of the materiality and management of something called "heritage". She suggests that recent perspectives coming from South and Southeast Asian scholarship on multi-religious sites, living heritage and challenges to monumentality provide a timely corrective.

Next, Indira Rajaraman examines the basis on which heritage sites might have been traditionally valued, and examines whether identification as a World Heritage Site preserves and builds further on that valuation, or destroys it. Although the notion of sustainability newly informs the discourse on heritage, it is largely restricted to concerns about the damage inflicted by increased tourist traffic. It is only when the basis of traditional valuation near these sites is understood that sustainability can be integrated into the discourse on World Heritage in a more informed way.

Using the example of the ancient city of Banavasi in modern Karnataka, Uthara Suvrathan explores the multiple layers that go into the making of historical sites. She moves from the academic, in this case archaeological, construction of the site as an important ancient capital city to examining the lived experience of the community and shows how, in many cases, the historical importance attributed by academics to Banavasi, is not always what is most important to people living within Banavasi. She also examines the present-day, locally initiated push towards cultural heritage tourism and its unclear and often uneasy relationship with academic approaches.

Moving from the global to the national, Part II includes chapters that discuss various facets of World Heritage Sites in India. These include caves at Ajanta, Ellora and Elephanta, the only three sites to be recognised in the state of Maharashtra, which could be classified as sites of both cultural and natural relevance; in fact, the monumentality of these rock-cut establishments is closely intertwined with their natural environments. By focusing on the caves at Ajanta, Elephanta and Ellora, Pia Brancaccio discusses the elusive boundaries between monuments and landscapes in South Asia, and how the dichotomy between cultural and natural heritage needs to be rethought in this context. The second chapter in this section, by Salila Kulshreshtha, on Nalanda Mahavihara, interrogates the continuation of the colonial legacy. Next, Swapna Liddle maintains that though the Qutub Minar complex was "monumentalised" in the 19th century and physically set apart from its surroundings, its history is closely tied to that of the neighbouring village, Mehrauli.

Part III shifts the discussion to transnationalism. Max Deeg examines the interpretation of the archaeological and historical evidence for the birthplace of the Buddha, identified with Lumbini (Terai, Nepal), since its inscription as World Heritage Site in 1997. He traces changes in the site after its inscription on the World Heritage List, through archaeological activities (conservation) and reconstruction of history (interpretation) in an attempt to promote particular ideological parameters. I. Wayan Ardika discusses challenges faced in preserving the sanctity of sites such as *Subak*, the Cultural Landscape of Bali Province inscribed in the World Heritage List in 2012. The Outstanding Universal Value of the site is based on *Tri Hita Karana*, which is a philosophical concept relating to harmony and balance between religion, society and the environment. Can this deeper understanding of the landscape be preserved against an onslaught from tourism ill-equipped to appreciate it? The last chapter, by Himanshu Prabha Ray, discusses India's transnational inscription of the Architectural Work of Le Corbusier in Chandigarh, Punjab in 2016. It raises several related issues: Is transnational nomination about inscription of as many sites as possible or is it about protection of universal values that would help focus on a shared heritage? It also raises the responsibility of maintaining equity in the inscription of transnational nominations among the different partners. Are UNESCO's guidelines for Outstanding Universal Value designed for promoting a worldview weighted towards powerful nation states? There are no clear answers to the question: Has inscription on the World Heritage List contributed to the goal of preservation of a plural heritage for posterity? Judging by some of the cases in India the way forward continues to be an uphill task. It is challenging, no doubt, but one which requires co-operation across several disciplines and several sectors.

Notes

1 Shashi Tharoor, 'The Need for a Museum on British Colonisation of India', 12 March 2017, www.aljazeera.com/indepth/opinion/2017/03/museum-british-colonisation-india-170312082632399.html (accessed on 7 July 2017).

2 No doubt there are studies that shift the focus to local agents and their participation in colonial knowledge production, such as Phillip B. Wagoner's argument ('Precolonial Intellectuals and the Production of Colonial Knowledge', *Comparative Studies in Society and History*, 2003, 45[4]: 783–814) that the military surveyor to the East India Company, Colin Mackenzie (1754–1821), used the services of local Niyogi Brahmanas to collect information on land revenue. The supposed 'objective' nature of the survey is an issue that was discussed by Peter Robb in 1998 (Peter Robb, 'Completing "Our Stock of Geography", or an Object "Still More Sublime": Colin Mackenzie's Survey of Mysore, 1799–1810', *Journal of the Royal Asiatic Society*, Third Series, 1998, 8[2]: 181–206), which Wagoner does not take into account nor the fact that Mackenzie's contribution lies in the official acceptance of the survey as a means of governance.

   Similarly, the edited book by Indira Sengupta and Daud Ali (*Knowledge Production, Pedagogy, and Institutions in Colonial India*, New York: Palgrave Macmillan, 2011) argues that "knowledge production, like the state project itself, was often more considerably fragmented, even while it operated within the framework of blatant and sustained asymmetrical power relations" (p. 4). This and other studies, however, do not address issues of implementation of the knowledge thus produced through legislation based on categories adopted and enforced by the colonial State.

3 James Fergusson's *History of Architecture* first appeared in 1855 as part of his well-known *Handbook*. A new edition very liberally enlarged appeared in 1862 also as part of a similar general *History of Architecture in all Countries*, while the third edition was published as *History of Indian and Eastern Architecture*, London: John Murray, 1876.

4 James Fergusson, *History of Indian and Eastern Architecture*, Vols. I and II, revised and edited by James Burgess (Indian Architecture) and R. Phene Spiers (Eastern Architecture), London: John Murray, 1910, pp. 4–5.

5 Fergusson, *History of Indian and Eastern Architecture*, pp. 39–40, 302.

6 Gary M. Tartakov, *The Durga Temple at Aihole: A Historiographical Study*, New Delhi: Oxford University Press, 1997, p. 6.

7 Himanshu Prabha Ray, 'From Multi-Religious Sites to Mono-Religious Monuments in South Asia: The Colonial Legacy of Heritage Management', in Patrick Daly and Tim Winter (eds.), *Routledge Handbook of Heritage in Asia*, London and New York: Routledge, 2012, pp. 69–84.

8 There are approximately 3,700 nationally protected monuments notified by the ASI. In the 1950s, a second category of 'state-protected monuments' was created in keeping with India's federal structure. For example,

of the total number of 1,200 structures of archaeological, historical and architectural importance in Delhi, only 174 are nationally protected, while another 33 were notified as state protected until 2012, though the final number may reach 92. Even then these numbers indicate only a miniscule protection for India's rich heritage.

9 Comptroller and Auditor General (CAG), 'Performance Audit of Preservation and Conservation of Monuments and Antiquities of Union Government', Report No. 18 of 2013, Ministry of Culture, p. 2.

10 This is an observation noted in the CAG, 'Performance Audit of Preservation and Conservation of Monuments and Antiquities of Union Government', p. 39.

11 Dom Moraes and Sarayu Srivatsa, *The Long Strider: How Thomas Coryat Walked From England to India in the Year 1613*, New Delhi: Penguin, 2003.

12 William Foster (ed.), *Early Travels in India 1583–1619*, London: Oxford University Press, 1921, p. 248.

13 Himanshu Prabha Ray, *The Return of the Buddha: Ancient Symbols for a New Nation*, London, New York and New Delhi: Routledge, 2014, pp. 98–133.

14 J. Ph. Vogel, 'Buddhist Sculptures From Benares', in *Archaeological Survey of India – Annual Report 1903–04*, New Delhi: Swati Publications, 1990 (reprint), pp. 222–223.

15 John S. Strong, *The Legend of King Aśoka*, Princeton, NJ: Princeton University Press, 1983, p. 19.

16 Strong, *The Legend of King Aśoka*, pp. 39, 39–70.

17 Volker Grabowsky, 'Buddhism, Power and Political Order in Pre-Twentieth Century Laos', in Ian Harris (ed.), *Buddhism, Power and Political Order*, London and New York: Routledge, 2007, p. 128.

18 Sir John Marshall and Alfred Foucher, *The Monuments of Sanchi*, texts of inscriptions edited, translated and annotated by N. G. Majumdar, 3 Vols., Calcutta and London: Probsthain, 1940, p. 12.

19 John Marshall, 'Where the Restored Relics of Buddha's Chief Disciples Originally Rested: The Stupa of Sariputta and Mahamogalana', *Illustrated London News*, 29 January 1949, p. 142.

20 Alexander Cunningham, *The Bhilsa Topes or the Buddhist Monuments of Central India*, London: Smith Elder and Company, 1854, p. 293.

21 Cunningham, *The Bhilsa Topes*, p. 308.

22 Cunningham, *The Bhilsa Topes*, pp. x–xi.

23 'Holy Buddhist Relic in Delhi', *The Times of India*, 24 May 1950, p. 6.

24 Saloni Mathur, *India by Design: Colonial History and Cultural Display*, Chicago: University of California Press, 2007, p. 150.

25 Julia Shaw, *Buddhist Landscapes in Central India: Sanchi Hill and Archaeologies of Religious and Social Change, c. Third Century B.C. to Fifth Century A.D.*, London: British Association for South Asian Studies, The British Academy, 2007.

26 Rani-ki-Vav (the Queen's Stepwell) at Patan, Gujarat, http://whc.unesco.org/en/list/922 (accessed on 8 July 2015).

27 Ankur Tewari, '5,000-Year-Old Harappan Stepwell Found in Kutch, Bigger Than Mohenjodaro's', *The Times of India*, 8 October 2014.

28 Kirit Mankodi, *The Queen's Stepwell at Patan*, Mumbai: Franco-Indian Pharmaceuticals Ltd., 1991, p. 30.

29 Purnima Mehta Bhatt, *Her Space, Her Story: Exploring the Stepwells of Gujarat*, New Delhi: Zubaan, 2014.
30 'A Walk Along Heritage and History', *The Hindu*, Thiruvananthapuram, 22 February 2016.
31 Jessica Moody, 'Heritage and History', in Emma Waterton and Steve Watson (eds.), *The Palgrave Handbook of Contemporary Heritage Research*, London: Palgrave Macmillan, 2015, pp. 113–129.
32 Lynn Meskell (ed.), *Global Heritage: A Reader*, Chichester: Wiley Blackwell, 2015, p. 1.
33 The following two have been influential in providing an intellectual basis for understanding the architectural history of the colonial period: Catherine B. Asher and Thomas R. Metcalf, *Perceptions of South Asia's Visual Past*, New Delhi: Oxford and IBH, 1994; Bernard S. Cohn, *Colonialism and Its Forms of Knowledge*, Princeton, NJ: Princeton University Press, 1996.
   The history of archaeology and the Archaeological Survey of India has been dealt with in the following studies: Dilip K. Chakrabarti, *A History of Indian Archaeology From the Beginning to 1947*, New Delhi: Munshiram Manoharlal, 2001; Upinder Singh, *The Discovery of Ancient India: Early Archaeologists and the Beginnings of Archaeology*, New Delhi: Permanent Black, 2004; Sudeshna Guha (ed.), *The Marshall Albums: Photography and Archaeology*, London: Alkazi Collection of Photography in Association with Mapin Publishing, 2010; Nayanjot Lahiri, *Marshalling the Past: Ancient India and Its Modern Histories*, Ranikhet: Permanent Black, 2012.
   Several practices relating to conservation and museums as practiced in independent India are discussed in Himanshu Prabha Ray, *Colonial Archaeology in South Asia (1944–1948): The Legacy of Sir Mortimer Wheeler in India*, New Delhi: Oxford University Press, 2007.
34 Bridget Allchin, F.R. Allchin and B. K. Thapar (eds.), *Conservation of the Indian Heritage*, New Delhi: Cosmo Publications, 1989.
35 F. R. Allchin, 'Threats to the Conservation of the Cultural Heritage in Rural and Urban Areas', in Bridget Allchin, F. R. Allchin and B. K. Thapar (eds.), *Conservation of the Indian Heritage*, New Delhi: Cosmo Publications, 1989, pp. 245–250.
36 Allchin, 'Threats to the Conservation of the Cultural Heritage', p. 248.
37 Tapati Guha-Thakurta, *Monuments, Objects, Histories: Institutions of Art in Colonial and Postcolonial India*, New Delhi: Permanent Black, 2004.
38 Alan Trevithick, 'British Archaeologists, Hindu Abbots, and Burmese Buddhists: The Mahabodhi Temple at Bodh Gaya, 1811–1877', *Modern Asian Studies*, July 1999, 33(3): 635–656; Alan Trevithick, *The Revival of Buddhist Pilgrimage at Bodh Gaya (1811–1949)*, New Delhi: Motilal Banarsidass, 2006.
39 Guha-Thakurta, *Monuments, Objects, Histories*, p. 61.
40 Deborah Sutton, 'Devotion, Antiquity, and Colonial Custody of the Hindu Temple in British India', *Modern Asian Studies*, January 2013, 47: 135–166.
41 Santhi Kavuri-Bauer, *Monumental Matters: The Power, Subjectivity, and Space of India's Mughal Architecture*, Durham, NC: Duke University Press, 2011.

42 Kavuri-Bauer, *Monumental Matters*, p. 2.
43 Kavuri-Bauer, *Monumental Matters*, p. 148.
44 Anne-Julie Etter, 'Antiquarian Knowledge and Preservation at the Beginning of the Nineteenth Century', in Indra Sengupta and Daud Ali (eds.), *Knowledge Production, Pedagogy and Institutions in Colonial India*, New York: Palgrave Macmillan, 2011, pp. 123–146.
45 Madhuri Desai, 'Resurrecting Banaras: Urban Space, Architecture and Religious Boundaries', PhD thesis, University of California, Berkeley, CA, 2007, pp. 282–283.
46 Vogel, 'Buddhist Sculptures From Benares', p. 74.
47 Jawhar Sircar, 'Mysteries and History – Where Did India Go Wrong With Its Museums and Monuments?', *The Telegraph*, Calcutta, Thursday, 10 December 2015.
48 *Asian Relations: Report of the Proceedings and Documentation of the First Asian Relations Conference, New Delhi, March–April 1947*, New Delhi: Asian Relations Organization, 1948.
49 'Delhi's 16th-century Purana Qila fort: a history of cities in 50 buildings, day 4', www.theguardian.com/cities/2015/mar/26/delhi-purana-qila-fort-history-cities-buildings (accessed on 5 February 2016).
50 This is an issue that has been discussed with reference to Buddhism in Ray, *The Return of the Buddha*.
51 *Resolutions adopted by the 75th Session of Indian History Congress, Mainstream*, Vol. LIII No 6, January 31, 2015 – Republic Day Special, http://www.mainstreamweekly.net/article5443.html (accessed on June 27 2018).
52 Giles Tillotson, *The Artificial Empire: The Indian Landscapes of William Hodges*, Richmond, VA: Curzon Press, 2000.
53 Farish Ahmad-Noor, 'An Unwilling Divorce: Colonial-Era Epistemology and the Division of South and Southeast Asia', in *Cultural and Civilisational Links Between India and Southeast Asia: Historical and Contemporary Dimensions*, New Delhi: Palgrave Macmillan, forthcoming.
54 Lynn Meskell, 'States of Conservation: Protection, Politics, and Pacting Within UNESCO's World Heritage Committee', *Anthropological Quarterly*, 2014, 87(1): 217–244.
55 http://mhrd.gov.in/international-cooperation-cell-4 (accessed on 5 October 2016).
56 Geri H. Malandra, *Unfolding a Mandala: The Buddhist Cave Temples at Ellora*, New York: State University of New York Press, 1993.
57 Lisa Owen, 'Absence and Presence: Worshipping the Jina at Ellora', in Himanshu Prabha Ray (ed.), *Archaeology and Text: The Temple in South Asia*, New Delhi: Oxford University Press, 2010.
58 Anna A. Suvorova, *Muslim Saints of South Asia: The Eleventh to Fifteenth Centuries*, London and New York: Routledge, 2004.
59 www.indiaculture.nic.in/project-mausam (accessed on 7 April 2017); Himanshu Prabha Ray (ed.), *Mausam: Maritime Cultural Landscapes Across the Indian Ocean*, New Delhi: National Monuments Authority and Aryan Books International, 2014.
60 Kumar Vikram, 'Project Mausam runs into Rough Weather', www.newindianexpress.com/thesundaystandard/2017/jan/21/project-mausam-runs-into-rough-weather-1562182.html (accessed on 7 April 2017); http://pib.nic.in/newsite/PrintRelease.aspx?relid=141133 (accessed on 7 April 2017).

References

Ahmad-Noor, Farish, 'An Unwilling Divorce: Colonial-Era Epistemology and the Division of South and Southeast Asia', in Shyam Saran (ed.), *Cultural and Civilisational Links Between India and Southeast Asia: Historical and Contemporary Dimensions*, New Delhi: Palgrave Macmillan, forthcoming.

Allchin, F. R., 'Threats to the Conservation of the Cultural Heritage in Rural and Urban Areas', in Bridget Allchin, F. R. Allchin and B. K. Thapar (eds), *Conservation of the Indian Heritage*, New Delhi: Cosmo Publications, 1989, pp. 245–250.

Asher, Catherine B., and Thomas R. Metcalf, *Perceptions of South Asia's Visual Past*, New Delhi: Oxford & IBH, 1994.

Asian Relations: Report of the Proceedings and Documentation of the First Asian Relations Conference, New Delhi, March–April 1947, New Delhi: Asian Relations Organization, 1948.

Bhatt, Purnima Mehta, *Her Space, Her Story: Exploring the Stepwells of Gujarat*, New Delhi: Zubaan, 2014.

Chakrabarti, Dilip K., *A History of Indian Archaeology From the Beginning to 1947*, New Delhi: Munshiram Manoharlal, 2001.

Cohn, Bernard S., *Colonialism and Its Forms of Knowledge*, Princeton, NJ: Princeton University Press, 1996.

Cunningham, Alexander, *The Bhilsa Topes or the Buddhist Monuments of Central India*, London: Smith Elder and Company, 1854.

Desai, Madhuri, 'Resurrecting Banaras: Urban Space, Architecture and Religious Boundaries', PhD thesis, University of California, Berkeley, CA, 2007.

Etter, Anne-Julie, 'Antiquarian Knowledge and Preservation at the Beginning of the Nineteenth Century', in Indra Sengupta and Daud Ali (eds), *Knowledge Production, Pedagogy and Institutions in Colonial India*, New York: Palgrave Macmillan, 2011, pp. 123–146.

Fergusson, James, *History of Indian and Eastern Architecture*, London: John Murray, 1876.

Fergusson, James, *History of Indian and Eastern Architecture*, Vol. I & II, Revised and edited by James Burgess (Indian Architecture) and R. Phene Spiers (Eastern Architecture), London: John Murray, 1910.

Foster, William (ed.), *Early Travels in India 1583–1619*, London: Oxford University Press, 1921.

Grabowsky, Volker, 'Buddhism, Power and Political Order in Pre-Twentieth Century Laos', in Ian Harris (ed.), *Buddhism, Power and Political Order*, London and New York: Routledge, 2007, pp. 121–142.

Guha, Sudeshna (ed.), *The Marshall Albums: Photography and Archaeology*, London: Alkazi Collection of Photography in Association with Mapin Publishing, 2010.

Guha-Thakurta, Tapati, *Monuments, Objects, Histories: Institutions of Art in Colonial and Postcolonial India*, Ranikhet: Permanent Black, 2004.

Kavuri-Bauer, Santhi, *Monumental Matters: The Power, Subjectivity, and Space of India's Mughal Architecture*, Durham, NC: Duke University Press, 2011.

Lahiri, Nayanjot, *Marshalling the Past: Ancient India and Its Modern Histories*, Ranikhet: Permanent Black, 2012.

Malandra, Geri H., *Unfolding a Mandala: The Buddhist Cave Temples at Ellora*, New York: State University of New York Press, 1993.

Mankodi, Kirit, *The Queen's Stepwell at Patan*, Mumbai: Franco-Indian Pharmaceuticals Ltd., 1991.

Marshall, John, 'Where the Restored Relics of Buddha's Chief Disciples Originally Rested: The Stupa of Sariputta and Mahamogalana', *Illustrated London News*, 29 January 1949, p. 142.

Marshall, John, and Alfred Foucher, *The Monuments of Sanchi*, The texts of inscriptions edited, translated and annotated by N. G. Majumdar, 3 Vols., Calcutta and London: Probsthain, 1940.

Mathur, Saloni, *India by Design: Colonial History and Cultural Display*, Chicago: University of California Press, 2007.

Meskell, Lynn, 'States of Conservation: Protection, Politics, and Pacting Within UNESCO's World Heritage Committee', *Anthropological Quarterly*, 2014, 87(1): 217–244.

Meskell, Lynn (ed.), *Global Heritage: A Reader*, Chichester: Wiley Blackwell, 2015.

Moody, Jessica, 'Heritage and History', in Emma Waterton and Steve Watson (eds), *The Palgrave Handbook of Contemporary Heritage Research*, London: Palgrave Macmillan, 2015, pp. 113–129.

Moraes, Dom, and Sarayu Srivatsa, *The Long Strider: How Thomas Coryat Walked From England to India in the Year 1613*, New Delhi: Penguin, 2003.

Owen, Lisa, 'Absence and Presence: Worshipping the Jina at Ellora', in Himanshu Prabha Ray (ed.), *Archaeology and Text: The Temple in South Asia*, New Delhi: Oxford University Press, 2010.

Ray, Himanshu Prabha, *Colonial Archaeology in South Asia (1944–1948): The Legacy of Sir Mortimer Wheeler in India*, New Delhi: Oxford University Press, 2007.

Ray, Himanshu Prabha, 'From Multi-Religious Sites to Mono-Religious Monuments in South Asia: The Colonial Legacy of Heritage Management', in Patrick Daly and Tim Winter (eds), *Routledge Handbook of Heritage in Asia*, London and New York: Routledge, 2012, pp. 69–84.

Ray, Himanshu Prabha, *The Return of the Buddha: Ancient Symbols for a New Nation*, London, New York and New Delhi: Routledge, 2014.

Ray, Himanshu Prabha (ed.), *Memory as History: The Legacy of Alexander in Asia*, New Delhi: Aryan Books International, 2007.

Ray, Himanshu Prabha (ed.), *Mausam: Maritime Cultural Landscapes Across the Indian Ocean*, New Delhi: National Monuments Authority and Aryan Books International, 2014.

'Report of the Comptroller and Auditor General of India', Report No. 18 of 2013, p. 39.

Robb, Peter, 'Completing "Our Stock of Geography", or an Object "Still More Sublime": Colin Mackenzie's Survey of Mysore, 1799–1810', *Journal of the Royal Asiatic Society*, Third Series, 1998, 8(2): 181–206.

Sengupta, Indra, and Daud Ali (eds), *Knowledge Production, Pedagogy, and Institutions in Colonial India*, New York: Palgrave Macmillan, 2011.

Shaw, Julia, *Buddhist Landscapes in Central India: Sanchi Hill and Archaeologies of Religious and Social Change, c. Third Century B.C. to Fifth Century A.D.*, London: British Association for South Asian Studies, The British Academy, 2007.

Singh, Upinder, *The Discovery of Ancient India: Early Archaeologists and the Beginnings of Archaeology*, New Delhi: Permanent Black, 2004.

Sircar, Jawhar, 'Mysteries and History – Where Did India Go Wrong With Its Museums and Monuments?', *The Telegraph*, Calcutta, Thursday, 10 December 2015.

Strong, John S., *The Legend of King Aśoka*, Princeton, NJ: Princeton University Press, 1984.

Sutton, Deborah, 'Devotion, Antiquity, and Colonial Custody of the Hindu Temple in British India', *Modern Asian Studies*, January 2013, 47: 135–166.

Suvorova, Anna A., *Muslim Saints of South Asia: The Eleventh to Fifteenth Centuries*, London and New York: Routledge, 2004.

Tartakov, Gary M., *The Durga Temple at Aihole: A Historiographical Study*, New Delhi: Oxford University Press, 1997.

Tewari, Ankur, '5,000-Year-Old Harappan Stepwell Found in Kutch, Bigger Than Mohenjodaro's', *The Times of India*, 8 October 2014.

Tharoor, Shashi, 'The Need for a Museum on British Colonization of India', 12 March 2017, www.aljazeera.com/indepth/opinion/2017/03/museum-british-colonisation-india-170312082632399.html (accessed on 7 July 2017).

Tillotson, Giles, *The Artificial Empire: The Indian Landscapes of William Hodges*, Richmond, VA: Curzon Press, 2000.

Trevithick, Alan, 'British Archaeologists, Hindu Abbots, and Burmese Buddhists: The Mahabodhi Temple at Bodh Gaya, 1811–1877', *Modern Asian Studies*, July 1999, 33(3): 635–656.

Trevithick, Alan, *The Revival of Buddhist Pilgrimage at Bodh Gaya (1811–1949)*, New Delhi: Motilal Banarsidass, 2006.

Vogel, J. Ph., 'Buddhist Sculptures From Benares', in *Archaeological Survey of India – Annual Report 1903–04*, New Delhi: Swati Publications, 1990 (reprint), pp. 222–223.

Wagoner, Phillip B., 'Precolonial Intellectuals and the Production of Colonial Knowledge', *Comparative Studies in Society and History*, 2003, 45(4): 783–814.

Part I
World Heritage Sites and cartographies

2 One world, two missions

UNESCO World Heritage in the making

Lynn Meskell

Introduction

Since the end of the Second World War the international community has sought to establish and maintain institutions that govern its common affairs. While such institutions take many forms, by far the most important have been formal international agreements through which nations bind themselves, under international law, to negotiated commitments.[1] The arena of global heritage is no different. That ideal of internationalism or world government found its fullest expression in the formation of the United Nations. This chapter examines UNESCO, the "visionary agency" of the United Nations, specifically its high-profile 1972 *Convention Concerning the Protection of the World Cultural and Natural Heritage*.[2] UNESCO itself represents the aspirations of an international community, the limitations of world government, concerns for protection and rights and notions of the global good. Yet the unresolved tensions that were present in the foundation of UNESCO in 1945 remain the same as those today confronting global governance of the world's natural and cultural heritage: cultural and religious difference, economic inequality, struggles for human rights, and fundamentally divergent understandings of the materiality and management of something called "heritage".[3]

From the outset UNESCO's institutional mandate has had two distinct strands with lasting legacies: one is the vision of fostering peace and multicultural dialogue broadly understood, while the other is an intergovernmental technocratic agency fostering "best practice".[4] Today the organisation still struggles how to balance these two core missions, technocracy and developing peace, as do its Member States in translating these priorities and practices on the ground. In this chapter I look at the recent membership of India in the World Heritage Committee and the presentation of its own national agenda, then examine how those international and national claims connect to high-profile World Heritage sites within India. In doing so I draw attention to the roles of other organisations and non-governmental organisations (NGOs) who, like government agencies, take heritage as foundational in the social, spiritual, economic and political lives of their citizens.

One worldism

UNESCO's visionaries attempted to reconcile the organisation's dual mission from the start. After the devastation of the Second World War, UNESCO's task was to promote peace and 'change the minds of men', primarily through fundamental education, celebration of cultural diversity and a dialogue between different peoples of the world. It considered its mission no less than steering the fate of civilisation itself. So on the one hand, UNESCO's first Director General, Julian Huxley, was a proponent of "scientific humanism" – a secular philosophy that advocates science, democracy and compassion over religious or supernatural belief.[5] On the other, powerful nations like the United States have always imputed that UNESCO should focus its intergovernmental efforts on the dissemination of technocratic expertise, rather than striving for international peace (an impossibility).[6] These two very differently perceived missions result in institutional schizophrenia and ineffectiveness that have garnered enduring international criticism. Today this is reflected in the current crisis UNESCO faces with the targeted destruction of World Heritage in the Middle East and other regions. For example, in recent calls to halt conflict, Western preservationist perspectives are still prioritised, ignoring other cultural practices and eliding the potential for a culturally diverse dialogue and inclusivity, so fundamental to UNESCO. Understandably, many onlookers perceive that a greater importance is being placed upon conserving monuments rather than saving lives, on technical preservation rather than wellbeing. Ultimately, the tensions that historically plagued UNESCO from its inception remain entrenched: science vs. religion, preservation vs. politics, peace building vs. technocracy.

Yet the extraordinary resilience of UNESCO's philosophy over some 70 years is noteworthy, crafted upon Western philosophical principles including scientific humanism, world citizenship and cultural progress. That history is itself instrumental in many aspects of World Heritage making and unmaking, from the increasing global desire to inscribe sites on the World Heritage List, to nations using World Heritage as an international tool for political and economic leverage, and finally, to the destruction of those sites for religious and political projects. UNESCO was formulated as an intergovernmental organisation founded after the ravages of the Second World War and was greatly influenced by its predecessor, the League of Nations.[7] Its mission to combat conflict, destruction and intolerance, while noble and much needed, has not been fully embraced, as recent conflicts over World Heritage sites in Yemen, Crimea and the Middle East have underlined.[8] Forging an international body with a mission for mutual cultural understanding through diversity on the one hand, and conservation of global patrimony through "technocratic hegemony" on the other has – at particular moments – engendered the very conflict it set out to dispel.

UNESCO's particular message of "world citizenship" through development was forged under liberal imperialism, offering the promise of a new scientific, modern, postnational future. Huxley masterminded a programme of cultural diversity and scientific modernity that was premised on the British civilising mission and social evolutionism.[9] Historian Mark Mazower claims that the founding of the United Nations itself was more about the endgame of empire, specifically the British Empire, rather than a product of American statecraft. British colonial efforts were thus bolstered by figures like Jan Smuts, Alfred Zimmern and Huxley himself.[10] UNESCO is a product of the mid 20th century, the prime example of a modernist institution, striving to overcome religious intolerance after the ravages of war in order to create a better future. Yet this was not, strictly speaking, a secular endeavour given UNESCO's physical setting in France and its cadre of European intellectuals who regarded the Bible as "the book" of UNESCO. Such implicit Eurocentrism continues to underwrite the intensifying struggles over religious practice and heritage, from UNESCO's great failure to save the Bamiyan Buddhas in Afghanistan, to the attacks on sites in Mali, Syria and Iraq. On a more quotidian level UNESCO has often struggled with the realities of living heritage sites, for example across much of South and Southeast Asia, particularly around daily issues of site use and access, and how these factors impact monumental conservation. The international, national and local understandings of heritage and monumentality are often at odds, and from the perspective India offers one vast continental context and challenge.

My research on UNESCO World Heritage employs a hybrid suite of methodologies including archival analysis, long-term participant observation and interviews, as well as drawing upon my own archaeological and ethnographic fieldwork from various global sites. Much of it could be described as an institutional ethnography, namely uncovering how "the social" is something that unites people's activities and how institutional organisations of power come to shape people experiences.[11] Through an ethnographic approach,[12] I hope to uncover new political dimensions of World Heritage that are typically occluded in complex international circuits. Specifically, my research narrows in on the politics of World Heritage properties and World Heritage Committee meetings, often captured in interviews with UNESCO ambassadors and diplomats, bureaucrats, heritage experts and many others who claim a local stake – all of which provide a vantage on the increasing desire and expansive investment towards a future in ruins. This work has taken me to UNESCO headquarters in Paris over many years but also entailed extended visits to India to talk to those in government, the Archaeological Survey of India (ASI), the Indian National Trust for Art and Cultural Heritage (INTACH) and other NGOs, heritage development agencies, consultants, academics, archaeologists and conservation architects. It is also noteworthy that these categories of experts are typically permeable, in that certain individuals often traverse many roles, adopting different personal stances and attachments, within the World Heritage arena.

India and World Heritage

From the outset India sought to play a strong role at the United Nations and to use the international arena for its own political projects. To his great credit, Jawaharlal Nehru demolished the last vestiges of imperial internationalism in a series of policy moves at the United Nations between 1946 and the mid-1950s. When considering the fate of South Africa and debating its Indian population, Nehru seized on the opportunity to bypass the British, using the UN platform for his own purposes of Indian self-determination and independence. Nehru argued that India was 'potentially a Great Power' and thus should not be 'treated like any small power'.[13] Using the newly established international organisation, India moved swiftly to the vanguard of the movement to challenge colonial domination and Nehru pushed home his advantage. Statesmen like Nehru have used UN agencies to critique colonialism, racism and European hegemony, calling for a new world order that would also prove expedient to domestic politics. India's role on the UNESCO World Heritage Committee, while more modest in aims and scope, has continued the powerful legacy of anti-colonial and anti-Eurocentric advocacy, forging South–South alliances and promoting national agendas on the global stage. This would complicate the standing of UNESCO's dual mission to promote world peace on one level, while also providing the highest levels of technical assistance on the other.

India ratified the World Heritage Convention in 1977 and has had three mandates to the World Heritage Committee (1985–1991, 2001–2007 and 2011–2015), the most recent of which forms part of the current study. As an intergovernmental agency and part of the UN family, States Parties like India that are signatories to the World Heritage Convention are the most powerful decision makers. For example, they decide whether a property is to be inscribed on the World Heritage List or placed on the World Heritage in Danger. The Committee is made up of 21 States Parties, elected at a General Assembly, that currently serve a four-year term. Yet the Committee's main representatives have shifted from being archaeological and environmental experts to almost exclusively state-appointed ambassadors and politicians. During India's last mandate to the Committee I tracked its participation and intervention during international meetings, led first by Ambassador Vinay Sheel Oberoi and then by Ambassador Ruchira Kamboj. I met with other members and experts from the Indian delegation in Paris and in Delhi, as well as with UNESCO representatives responsible for South Asian heritage. I interviewed government officials responsible for managing World Heritage sites, as well as many other experts involved in heritage consultancies, NGOs and civil society. To date, India has 27 cultural sites inscribed on the World Heritage List, seven natural and one mixed site. During my fieldwork I conducted site visits and interviews around the monuments of Hampi, Taj Mahal and Fatepur Sikri, Red Fort, Humayan's Tomb and Qutub Minar in Delhi, Jantar Mantar in Jaipur, the Western Ghats and several of the hill forts in Rajasthan. Specifically, this enabled me to examine if the official position of the national delegation reflects, constitutes or contradicts what happens at World Heritage sites, comparing how international positioning of specific sites connects to heritage practices on the ground.

Apart from its domestic agendas, India, like most States Parties, has advanced its international ambitions through its membership of the World Heritage Committee. During its last mandate (2011–2015) India worked in close partnership with other nations of the BRICS group, a formidable geo-political alliance between Brazil, Russia, India, China and South Africa.[14] The acronym BRICS was coined at Goldman Sachs for those nations at a similar stage of newly advanced economic development and the subsequent shift in global economic power away from the older-styled developed G8 countries. Over the past several years all five BRICS members have served on the World Heritage Committee. During that time they were increasingly outspoken in their dissatisfaction with the workings of the World Heritage Secretariat and the Advisory Bodies,[15] they questioned the Convention's Operational Guidelines and what constitutes expertise and technical authority.

For example, throughout Committee deliberations over controversial damming in the Portuguese Alto Douro Wine Region in 2012, Ambassador Oberoi asserted that the 'pyramids would never have been built if ICOMOS and the World Heritage Committee had been there'.[16] ICOMOS (International Council on Monuments and Sites), he imputed, did not stand for heritage, and was constantly overstepping the mark. Taking the lead from India, direct challenges like these to UNESCO's standing and mission have enabled more and more States Parties to intervene, bolstered by the critique that the List is biased towards European properties and must be more balanced. While this critique is certainly valid, the rise of BRICS has also enabled more overt political lobbying and further attacks on the World Heritage system and its particular forms of technical expertise.[17] India's stated challenge to Eurocentric priorities, values and conservation techniques may appear to be a much-needed, possibly even post-colonial critique. Yet one cannot be sure whether the same position is shared by other BRICS nations like China or the Russian Federation. In reality, many of the positions adopted and speeches given by States Parties on the Committee are pre-ordained by their own central governments,[18] and often are more about achieving national aims and bolstering international alliances than the protection of heritage sites.

India's critique of technical expertise, particularly in regard to the workings of ICOMOS, has masked a broader political agenda to support other BRICS nations, as in the problematic conservations cases of Mapungubwe Cultural Landscape[19] and Bolgar Historical and Archaeological Complex,[20] guaranteeing that infrastructure, inappropriate reconstruction or development projects could continue unhindered. India's support of South Africa ensured that construction of a coal mine in the Mapungubwe's buffer zone would not impinge on its World Heritage listing (see Figure 2.1).[21] In fact, India claimed that there would be no impact from the open-cast mine and, when a less-destructive alternative was proposed, it countered that the State Party should choose the mining technology. The Committee cannot comment on "technical issues", Ambassador Oberoi contended, since "we are not expert".

In concert with other BRICS members, India vocally supported the construction of a vast maritime highway encircling the historic district of Panamá,[22] ignoring threats to the historic fabric of the settlement and adverse impacts upon its residents.[23] India argued that governments, as sovereign entities, should not be put on the line, nor should information from NGOs and other groups necessarily be accepted. The main criticism voiced over India's four-year term could be summarised

Figure 2.1 The Indian Delegation at the World Heritage Meetings in Phnom Penh, Cambodia 2013.

Source: Official UNESCO photo taken by Sin Sareth

as a challenge to notions of Western expertise, a reaffirmation of local knowledge and cultural understandings, and an emphasis on vernacular and living heritage. India's role in the Committee underscores the mounting tension towards UNESCO's mission and purpose more generally, advocating cultural specificity in the heritage sphere and promoting development, but also challenging what constitutes technical expertise. While many of India's interventions found resonance with other States Parties I interviewed, particularly those from the global South, they also effectively served India's national interests and presented a picture of its own vibrant heritage scene and homegrown expertise.

Over the years, members of the Indian delegation have recounted experiences to me that suggest India has substantive grounds to be critical of the World Heritage process. When nominating their own properties – like the well-preserved and documented Hill Forts of Rajasthan[24] – evaluators were chosen by ICOMOS with no regional or site-specific expertise and could not communicate in English. This same criticism is repeatedly made by many States Parties, as is the concern that the pool of "expert" evaluators is limited so that the same individuals are chosen repeatedly. Additionally, several Indian delegates felt that wealthy European nations like Italy, Germany and France simply pay for experienced consultants to put the necessary spin on dossiers that appeal to the Advisory Bodies and, increasingly, the same individuals are involved at various levels of the process, leading to charges of conflict of interest. As such, there have been calls for greater transparency: more than one official suggested that expert evaluators be named. This has become a common request. Others expressed the view that European countries that have dominated the List should withhold nominations, allow for a more representative and balanced List.[25]

Discussing the importance of World Heritage in Delhi, several diplomats who have served on the Indian delegation over the years described World Heritage status as primarily important in emotive terms. The reasons for this were international recognition, rather than generating socio-economic revenue, despite the visibility of state marketing campaigns like *Incredible India*. Another irony is that back in the 1960s it was India that had approached UNESCO to develop the first cultural tourism development plans.[26] Those I spoke to in the Ministry of Culture certainly felt that visibility, status and tourism were strongly pegged to the UNESCO brand and were keen to expand these connections further. Despite India's push to have more and more sites inscribed on the World Heritage List, proudly claiming that it had secured three sites (not the limit of two) in 2016, various officials argued that intangible heritage was more important and that communities prefer living heritage, instead of the material sites.

While the intangible aspects of culture are clearly vital in the South Asian context, there still remains the pressing issue of multicultural histories and traditions at World Heritage sites. Reflecting multiple occupations by various cultures and religions through time is an ongoing concern primarily for scholars and experts, expressed during many interviews in India, rather than government officials. As Himanshu Prabha Ray has effectively demonstrated, many major archaeological sites in India bear witness to an intense layering of different cultures, religions and histories. These experiences can be anchored in places or traced across landscapes and seascapes and not simply instilled in architecture. Instead of perpetuating a colonial monument-centered or isolationist view, she considers it more culturally appropriate to move away from narrow site-based studies towards more expansive cultural landscapes. Going further, and in light of this cultural pluralism, she suggests that transnational nominations to the World Heritage List might also be more fitting in the South Asian context.[27] Her own work on the potential nomination of Mausam, linking numerous historic sites and heritage practices across the Indian Ocean in a vast international seascape, provides a compelling example.[28] This would accord well with UNESCO's own mission for cultural exchange and dialogue between peoples.

On the face of it, the ASI manages India's cultural World Heritage. This organisation was set up by the British in 1861, with Alexander Cunningham its first Director General. Back in the 1950s Mortimer Wheeler claimed that although it was 100% Indian, its inadequacy owed its inheritance to British direction. He also regarded it as the 'largest and most complex archaeological machine in the world'.[29]

The scale of heritage protection in India is indeed daunting, the ASI having neither adequate resources nor the funding to manage, maintain, restore or fully develop India's enormous cultural patrimony. Currently there are more than 3,650 monuments on the National Heritage register, as well as thousands of sites that remain unlisted and ostensibly unprotected. For instance, Delhi has 174 sites on the national list according to INTACH,[30] whereas they have 1200 informal heritage sites registered. Practitioners from the NGO have been attempted to challenge Eurocentric modes of monumentality, conservation and site use.[31] The ASI is frequently charged with maintaining a monument-centered approach stemming from its colonial legacy.[32] Most of the designated religious structures, according to A. Krishna,[33] though still affiliated with the various faiths, are not in active use or no longer support worship or liturgical practices. It is noteworthy too that the majority of sites put forward for inscription on the World Heritage List, apart from being managed by the ASI, were those sites celebrated and conserved by British colonial figures like Cunningham, Curzon and Marshall: the Taj Mahal,[34] Fatehpur Sikri, Sarnath, Konark, Sanchi, Delhi's Red Fort, the Chola temples and so on. Given India's history of anti-colonial resistance and celebration of its distinct practices, it is striking that its selection for UNESCO listing has mirrored British colonial priorities. It should also be said that earlier excavation, conservation and restoration measures under colonial rule remain deeply problematic, an unfortunate situation that has been prolonged under ASI management. One need only examine colonial and later conservation strategies at the World Heritage site of Humayan's Tomb in comparison with more recent work undertaken by the Aga Khan Trust for Culture (AKTC).[35]

While the ASI has received criticism from today's heritage and conservation lobbies, it has at times also alienated its own local

Figure 2.2 Demolished buildings in the bazaar in front of the Virupaksha
temple, Hampi World Heritage Site, August 2016.

Source: Photo by the author

communities by forcibly removing people from in and around sites,
severing connections of place and prohibiting religious practices (see
Figure 2.2). For instance, scholars have noted the growing dissatisfaction of local residents around the Taj Mahal in Agra in terms of
escalating religious tensions between different groups, the creation
of slums, lack of benefits from tourism, and alienation of locals in
favour of tourists and their experiences.[36] In the case of Hampi World
Heritage site in Karnataka, the ASI was responsible for destroying
property, forcing hundreds of local residents to lose their homes and
livelihoods, leaving in their wake destroyed buildings and exposed
excavation trenches near the Virupaksha temple (see Figure 2.3).
When speaking to former Hampi residents about the destruction and
relocation, it was evident that many other intrusive structures across
the site have remained intact, revealing the contradictions, inconsist-
encies and ultimately potential corruption, all of which combine to
create a sense of embittered alienation.[37] Taken together this national

Figure 2.3 Excavation trenches near the Virupaksha temple, Hampi World Heritage site, August 2016.

Source: Photo by the author

agency has been amply critiqued by numerous Indian scholars themselves in terms of both their technical expertise as well as the ASI's commitment to living communities, the two principles to which UNESCO claims to be wed.

Yet the story is more complex, since the ASI is directly under the control of the Ministry of Culture and can itself exercise only limited autonomy. The Ministry selects which sites are nominated to the World Heritage List and it is responsible for their long-term management and development. However, the Indian government's independent audit in 2013 of the preservation and conservation of monuments found that World Heritage sites did not receive appropriate care and protection, there was unauthorised construction in and around certain sites and no comprehensive assessment of preservation had ever been carried out. Ultimately it was the Ministry's governance that was deemed deficient on 'aspects of adequacy of policy and legislation, financial management, monitoring of conservation projects'.[38] D.K. Chakrabarti argues that various tasks at the ASI are routinely outsourced: a finding reiterated by the 2013 audit.[39] In terms of World Heritage dossiers, for example, the auditors found that the government was increasingly outsourcing the preparation of dossiers to external consultants. With the increased use of consultants, they noted a steady decline in acceptance of proposals to UNESCO's List, despite significant consultancy fees. The auditor's 'scrutiny revealed lack of transparency, tendering irregularities and undue favours to consultants'.[40] Moreover, even if a dossier was well prepared and the site inscribed, World Heritage status did not translate into better availability of facilities, funding and staffing. Major conservation problems were also documented for UNESCO properties like the Taj Mahal, Red Fort and Ajanta caves.

The foregoing already suggests a major disjuncture between the official face of Indian heritage ideologies and practices presented in UNESCO's international arena as opposed to the raft of challenges that the state must deflect domestically. Here I suggest that the official positions advocated by the State Party in World Heritage Committee meetings, not simply about Indian sites but also in relation to other properties, described and defended at the highest level, are significantly at odds with the national situation on the ground. Paying particular to local contexts, I suggest that while there are noteworthy Indian initiatives supporting living heritage, multiculturalism and humanitarian efforts, these have almost entirely developed through networks of NGOs, heritage consultancies and civil society.

Heritage alternatives

I want to briefly mention three different initiatives, each underscoring the potentials for developing Indian World Heritage Sites. All three intersect with UNESCO's stated philosophies in different ways and are the work of non-state agencies. In the case of Jantar Mantar, Jaipur, I cite initiatives by the Development and Research Organisation for Nature, Arts and Heritage (DRONAH) as an example of site development and presentation that aims for community revitalisation. With Humayan's Tomb in New Delhi, it is both the advanced conservation strategies and many community uplift projects of the AKTC that warrant attention. Finally, around the Taj Mahal in Agra, I describe work by the Center for Urban and Regional Excellence (CURE), a non-profit dedicated to developing sustainable heritage tourism that benefits communities while also delivering clean water, sanitation and infrastructure. While such projects are progressive, I am also reminded of Nakamura's critique that developmental tropes including "participatory planning", "sustainable development", and "capacity building" have entered into the heritage lexicon in India and that these strategies are often employed to engage local, often impoverished stakeholders. In so doing, she argues, there always exists a danger in bartering access to neoliberal pathways and resources in exchange for the rights to the past.[41]

Jaipur's Jantar Mantar is an astronomical observation site built in the early 18th century with a set of some 20 fixed instruments. The site represents the most significant, comprehensive and the best preserved of India's historic observatories (see Figure 2.4).[42] DRONAH produced the nomination dossier for UNESCO listing, in which they offered a sustainable and integrated development of the site, management plan and conservation strategy. This included a number of secondary plans including an environmental and landscape plan, a comprehensive mobility plan and interpretation and visitor management plan for the site. Inscribed in 2010, Ahmed claims it is one of the few Indian World Heritage sites where the management plan for the site is integrated into the city's master plan.[43] The site is further connected to wider initiatives across Jaipur to revive craft and foster local development through heritage walks. Here the specific aims are to improve conditions for local craftspeople, to support local businesses and develop infrastructure projects for improved drainage, solid waste management or rainwater harvesting with participation of the residents along the walk.[44] DRONAH Foundation,[45] the associated non-profit, has as its mission to promote and encourage environmental rehabilitation and human health, a participatory approach to built heritage and promoting indigenous techniques and crafts in the area of living heritage.

Figure 2.4 Jantar Mantar World Heritage site, Jaipur, August 2016.

Source: Photo by the author

Jantar Mantar was also the first cultural World Heritage site in India protected by a State Department of Archaeology. Moreover, the Government of Rajasthan passed an order in 2011 that two-thirds of the site revenues should go to the Rajasthan Heritage Development Management Authority (RHDMA) for regular maintenance and management of the property, thus ensuring the long-term maintenance of Jantar Mantar. Apart from the professional presentation of the site and information provided there is an impressive interpretation centre with state-of-the-art displays, reconstructions and explanatory videos. From this perspective Jantar Mantar differs from most World Heritage sites in India. Its success suggests a model that might be followed at other historic and archaeological sites. It is also an example where various agencies and individuals collaborate to produce forms of heritage divergent from national models.

Another successful programme that specifically targets both conservation and community uplift is AKTC's work at Humayan's Tomb. Constructed in 1570, this was the first garden-tomb on the Indian subcontinent and inspired several major architectural innovations, culminating in the construction of the Taj Mahal.[46] Humayan's Tomb was inscribed on the World Heritage List in 1993 (see Figure 2.5). In 1997, to celebrate the 50th anniversary of India's independence, the Aga Khan offered to restore the garden of Humayun's Tomb. In 2007, working with the ASI, the Municipal Corporation of Delhi and the Central Public Works Department, AKTC began a broader urban renewal initiative. Since then it has been involved in the conservation of more than 30 monuments, the creation of a 100-acre city park by landscaping the Sundar Nursery – Batashewala Complex and significant improvements to the quality of life for the residents of Hazrat Nizamuddin Basti.[47] The organisation supports a traditional craft-based approach to the conservation of India's monumental buildings, with more than 200,000 man-days of employment accomplished by master craftspeople to date. On the one hand it claims to be the only private agency in India working with monuments that similarly offers a model for civil society engagement in urban development. On the other, AKTC's understanding of craft traditions is based on exhaustive documentation using technology such as 3D laser scanning, archival research, peer review by independent national and international experts, and high levels of supervision.[48] In 2016 it was responsible for successfully extending the boundaries of the World Heritage site, providing conservation outreach programmes and initiating an interpretation centre.[49]

More than any other agency focused on conservation I encountered in India, AKTC has the most social programmes that positively impact

Figure 2.5 AKTC's work in progress at Humayun's Tomb World Heritage site.
Source: Photo by Narendra Swain and courtesy of the AKTC.

local communities. Describing itself as a non-profit People Public-Private Partnership, AKTC has invested in childhood and women's health and educational schemes (child care, computer access, school management, clinical care, health programmes), cultural revival at Nizamuddin, rehabilitation of civic spaces (resource centres), environmental projects (desilting, landscaping, park creation), improving sanitation (building toilets, solid waste management schemes) and so on. Many of its projects interweave living heritage and religious practices around historic sites, without privileging the monumental approach for which UNESCO is so often critiqued. The work of AKTC at Humayan's Tomb embodies the perspective advocated by the Indian delegation over its years on the World Heritage Committee. While the AKTC works loosely with the ASI by virtue of national management of UNESCO sites, it works with numerous other partners, from TATA Trust to the Norwegian Ministry of Foreign Affairs. Given the scale and complexity of this project, such collaborations offer an effective model. Lastly, it should be stated that AKTC's approach to the site of Humayan's Tomb also reflects UNESCO's two missions: technical proficiency in the sphere of conservation and development that benefits future generations.

Finally, the two UNESCO World Heritage properties in Agra are the Taj Mahal[50] and the nearby Red Fort of Agra,[51] both inscribed in 1983. While CURE is a non-profit concerned more with community participation and benefits within urban development than heritage conservation per se, its projects often combine historic places and living communities. Some of its team trained as conservation architects. For instance, in Agra CURE developed heritage walks that, since 2006, have been run by trained local youths. The Mughal Heritage Walk links the ancient village of Kucchpura across from the Taj Mahal, to the first observatory built by Shahjahan, the Humayun Mosque that pre-dates the Taj and the Mehtab Bagh or the Black Taj site. The Taj Culture Walk ventures into the city's heart, Tajganj, to see the arts and crafts of Mughal India (marble inlay work, zardozi embroidery, pottery making, kite and pigeon flying contests). Not only do ticket sales go towards community development projects, but also CURE has initiated a number of projects with international partners.[52] This includes restoring wells, promoting water conservation, harvesting rainwater, recharging groundwater, and treating and reusing wastewater. CURE has three primary goals: to reconnect urban societies and enable decision making to ensure sustainable development; to strengthen local agencies for participatory community development; and to build ground-up, effective service delivery.

In Agra CURE is dedicated to low-income communities, and the people I interviewed described its efforts as contributing to programmes, practice and policy. One leading figure spoke at length about both the 'tyranny of participation' and the 'paradox of empowerment'. As such CURE is conscious to employ approaches that are resolutely bottom up. For example, after complaints about tourist experiences in Agra it worked with local businesses and tour operators, as well as youth groups, to overcome the culture of alienation felt by the community and to ensure that benefits were shared. Another primary concern is lack of sanitation. CURE has built toilets and supported a community-operated Decentralised Waste Water Treatment System (DEWATS). According to many of the experts I talked to in cities like Agra, Delhi and Jaipur, it is not possible to entertain the luxury of heritage conservation without attending to the basic needs of sanitation, water, safety and livelihoods. These are even more pressing issues for women in terms of their day-to-day safety, health and wellbeing. Gender is a major consideration in many of the social programmes I witnessed. It is perhaps no coincidence that many of the most active and inspirational individuals working in these organisations today are women.

Conclusions

Contemporary global struggles for preserving the past are increasingly caught between the social, spiritual, economic and political needs of the present and particular attachments to the material world. UNESCO's 1972 World Heritage Convention is one of the most high-profile arenas where those tensions unfold today. This chapter has considered the organisation's dual mission, with particular attention to the Indian setting. Attempting to disseminate Western technocratic models worldwide, on the one hand, and prescribing the route to peace on the other, extends UNESCO's utopian 20th century dream into the 21st. However, many local and indigenous communities, whether in Europe or Asia, are calling for a withdrawal of their UNESCO status, complaining that the burdens of conservation and impositions of their governments impinge too heavily upon their lives. At a time when UNESCO entwines culture and sustainable development in new programmes and missions, many communities remain vulnerable to the translation of these proscriptions by their own governments. World Heritage recognition has also provided the rationale for violence, forced removal and curtailing of rights as witnessed in sites as diverse as Hampi, Panama City, Petra and Angkor. UNESCO recognition, as implemented by the nation state, may not always deliver positive benefits. Moreover, "living heritage" as a concept and cultural tradition is something that many of my interlocutors would argue World Heritage has yet to fully embrace or integrate within the workings of the 1972 Convention. Yet my preliminary fieldwork in India shows that there are alternatives at the local level to living with heritage and that the greatest innovation is being driven by civil society, by nongovernmental, non-profit organisations. While there will inevitably be critique and shortcomings, groups like INTACH, CURE, AKTC and DRONAH Foundation are attempting to improve the lives of connected communities, often addressing the most basic human needs rather than simply focusing on preservation efforts and other forms of "fortress conservation".[53]

Many experts interviewed during my time in India argue that the call to embrace Asian perspectives and living heritage modalities that India has so often promulgated at UNESCO World Heritage Committee meetings is not actually implemented or institutionalised by the government itself. As the national authority, the ASI remains wed to older forms of colonial conservation and management, often positioning people as the problem. As one expert from Rajasthan explained, this is not only an issue of scale – the vast number of sites, their size and complexity – but also embracing and acknowledging the human scale of maintenance, visitation and use. Official heritage work in India remains 'beholden to protecting monumental materialities and state and national agendas rather than attending to the unique historicity of its monuments (e.g. caves and forts) and monumental urban forms (e.g. settler cities)'.[54] Perhaps not surprisingly the more progressive perspectives and projects are coming from NGOs, foundations and even heritage consultancies. On a smaller scale, the projects briefly outlined here are attempting to inculcate progressive conservation practice, while being attendant to social uplift and concrete local benefits. Taken together, these organisations, while often on the fringes of UNESCO's institutional processes, actually attempt to bridge the dual mission that UNESCO's founders sought to implement worldwide.

In ending, people often ask me, is the world a better place because of UNESCO? Personally I think yes. Is there something to be salvaged from the original vision of men like Huxley, or was UNESCO destined to unfold as a colonising, and now globalising, set of practices inflected with the inherently Western technocratic and civilising mission? That is a more difficult proposition. But this remains a critical moment to ask some uncomfortable questions about the entanglements and obligations between international agencies, nations and their citizens, in order to save the past, in the present, for the future.

Acknowledgements

My greatest debt is to Himanshu Prabha Ray for her inspiration and guidance over the last few years. She first invited me to India in 2013 and has supported my nascent research there ever since. I also want to extend my gratitude to the many experts and scholars who have been generous with their time, ideas and contacts including Pooja Agrawal, Amita Baig, Hemani Baydal, Moe Chiba, John Fritz, Divay Gupta, Rima Hooja, Amrit Jha, Shikha Jain, Renu Khosla, Nayanjot Lahiri, Swapna Liddle, AGK Menon, Ratish Nanda, Vinay Sheel Oberoi, Nupur Prothi, Indira Rajaraman, Jahnwij Sharma, Ravindra Singh, Nalini Thakur and Mudit Trivedi. I would also like to thank Claudia Liuzza and Carolyn Nakamura for their long-term collaboration and friendship. My initial fieldwork was supported under the Australian Research Council's Discovery scheme (The Crisis in International Heritage Conservation in an Age of Shifting Global Power – DP140102991). I am also grateful to the GIAN programme and the many scholars and students I had the pleasure of interacting with at Jawaharlal Nehru University.

Notes

1 G. Sluga, 'Imagining Internationalism', *Arts: The Journal of the Sydney University Arts Association*, 2012: 32; *Internationalism in the Age of Nationalism*, Philadelphia, PA: University of Pennsylvania Press, 2013.
2 http://whc.unesco.org/en/convention/ (accessed on 20 December 2016). The Convention created a set of obligations to protect the past for future generations, an aspiration for a shared sense of belonging, and a global solidarity, see F. Choay, *The Invention of the Historic Monument*, Cambridge: Cambridge University Press, 2001, p. 140.
3 L. M. Meskell, 'States of Conservation: Protection, Politics and Pacting Within UNESCO's World Heritage Committee', *Anthropological Quarterly*, 2014, 87(1); Lynn Meskell, 'Gridlock: UNESCO, Global Conflict and Failed Ambitions', *World Archaeology*, 2015, 47(2): 225–238; L. M. Meskell et al., 'Multilateralism and UNESCO World Heritage: Decision-Making, States Parties and Political Processes', *International Journal of Heritage Studies*, 2015, 21(5).
4 V. Pavone, 'From Intergovernmental to Global: UNESCO's Response to Globalization', *The Review of International Organizations*, 2007, 2(1); V. Pavone, *From the Labyrinth of the World to the Paradise of the Heart: Science and Humanism in UNESCO's Approach to Globalization*, New York: Lexington, 2008.
5 R. S. Deese, *We Are Amphibians: Julian and Aldous Huxley on the Future of Our Species*, Berkeley, CA: University of California Press, 2014; G. Sluga, 'UNESCO and the (One) World of Julian Huxley', *Journal of World History*, 2010, 21(3).

6 Pavone, 'From Intergovernmental to Global: UNESCO's Response to Globalization', pp. 80–81.

7 Before UNESCO there was another intergovernmental agency, forged at the end of a world war in order to promote peace, namely the League of Nations (1920–1946). The League had grand ambitions for cultural sites around the world, particularly in fostering research into ancient civilizations, the protection of sites and the establishment of museum collections. While much was made of its international representation in archaeology, the truth was rather more a European and North American affair with scant, if any, mention of scholars from Africa, Latin America, Asia and the Pacific or concern for their heritage.

8 L. M. Meskell, 'World Heritage and WikiLeaks: Territory, Trade and Temples on the Thai-Cambodian Border', *Current Anthropology*, 2016, 57(1); C. De Cesari, 'World Heritage and Mosaic Universalism', *Journal of Social Archaeology*, 2010, 10(3); 'World Heritage and the Nation-State', in C. De Cesari and A. Rigney (eds.), *Transnational Memory: Circulation, Articulation, Scales*, Berlin: de Gruyter, 2014; C. Joy, *The Politics of Heritage Management in Mali: From UNESCO to Djenné*, Walnut Creek, CA: Left Coast Press, Inc., 2012; ' "UNESCO Is What?" World Heritage, Militant Islam and the Search for a Common Humanity in Mali', in C. Brumann and D. Berliner (eds.), *World Heritage on the Ground: Ethnographic Perspectives*, Oxford: Berghan, 2016.

9 P. Betts, 'The Warden of World Heritage: UNESCO and the Rescue of the Nubian Monuments', *Past & Present*, 2015, 226 (suppl 10); 'Humanity's New Heritage: UNESCO and the Rewriting of World History', *Past & Present*, 2015, 228(1); J. Toye and R. Toye, 'One World, Two Cultures? Alfred Zimmern, Julian Huxley and the Ideological Origins of UNESCO', *History*, 2010, 95(319); G. Sluga, 'Editorial – the Transnational History of International Institutions', *Journal of Global History*, 2011, 6(2); 'Imagining Internationalism'; *Internationalism in the Age of Nationalism*.

10 M. Mazower, *No Enchanted Palace: The End of Empire and the Ideological Origins of the United Nations*, Princeton, NJ: Princeton University Press, 2009; 'The End of Eurocentrism', *Critical Inquiry*, 2014, 40(4).

11 B. Müller, 'Lifting the Veil of Harmony: Anthropologists Approach International Organizations', in B. Müller (ed.), *The Gloss of Harmony: The Politics of Policy-Making in Multilateral Organisations*, London: Pluto Press, 2013; R. Bendix, 'The Power of Perseverance: Exploring the Negotiation Dynamics at the World Intellectual Property Organization', in B. Müller (ed.), *The Gloss of Harmony: The Politics of Policy-Making in Multilateral Organisations*, London: Pluto Press, 2013; C. Shore, 'European Integration in Anthropological Perspective: Studying the "Culture" of the EU Civil Service', in R. A. W. Rhodes, P. t' Hart and M. Noordegraaf (eds.), *Observing Government Elites: Up Close and Personal*, New York: Palgrave Macmillan, 2007.

12 L. M. Meskell, 'Archaeological Ethnography: Conversations Around Kruger National Park', *Archaeologies: Journal of the World Archaeology Congress*, 2005, 1(1); *The Nature of Heritage: The New South Africa*, Oxford: Wiley Blackwell, 2011.

13 Mazower, *No Enchanted Palace: The End of Empire and the Ideological Origins of the United Nations*, p. 170.

14 See also R. Hoggart, *An Idea and Its Servants: UNESCO From Within*, Piscataway, NJ: Transaction, 2011; I. B. Claudi, 'The New Kids on the Block: BRICs in the World Heritage Committee', MA Thesis, Department of Political Science, University of Oslo, Norway, 2011.

15 The Advisory Bodies are comprised of international experts who conduct monitoring missions and evaluations: ICCROM, ICOMOS and the International Union for Conservation of Nature (IUCN). Outspoken Brazilian and Indian delegates during the 2013 meetings suggested changing the Operational Guidelines, bringing in independent evaluators, reducing the role of the Advisory Bodies, inscribing a larger quota of sites per country and so on.

16 Cited in Meskell et al., 'Multilateralism and UNESCO World Heritage: Decision-Making, States Parties and Political Processes', p. 427. ICOMOS was founded in 1965 and is an international NGO. It provides evaluations of cultural properties including cultural landscapes proposed for inscription on the World Heritage List.

17 Meskell, 'States of Conservation: Protection, Politics and Pacting Within UNESCO's World Heritage Committee'.

18 Individual speakers within national delegations can take a more independent stance and India is well-known for having informed and articulate members. However, generally on important decisions, States Parties increasingly receive instructions from their capitals.

19 http://whc.unesco.org/en/list/1099

20 http://whc.unesco.org/en/list/981 For analysis see G. Plets, 'Ethno-Nationalism, Asymmetric Federalism and Soviet Perceptions of the Past: (World) Heritage Activism in the Russian Federation', *Journal of Social Archaeology*, 2015, 15(1); Meskell et al., 'Multilateralism and UNESCO World Heritage: Decision-Making, States Parties and Political Processes'; Bertacchini, Liuzza and Meskell, 'Shifting the Balance of Power in the UNESCO World Heritage Committee: An Empirical Assessment', *International Journal of Cultural Policy*, 2015, 23(3): 331–351; Bertacchini et al., 'The Politicization of UNESCO World Heritage Decision Making', *Public Choice*, 2016, 167(1–2): 95–129.

21 L. M. Meskell, 'From Paris to Pontdrift: UNESCO Meetings, Mapungubwe and Mining', *South African Archaeological Bulletin*, 2011, 66(194).

22 http://whc.unesco.org/en/list/790

23 Meskell, 'States of Conservation: Protection, Politics and Pacting Within UNESCO's World Heritage Committee'.

24 http://whc.unesco.org/en/list/247. See S. Jain and R. Hooja, 'The Hill Forts of Rajasthan as a Serial Nomination', *Context*, 2013, X(2).

25 See S. Labadi, 'A Review of the Global Strategy for a Balanced, Representative and Credible World Heritage List 1994–2004', *Conservation and Management of Archaeological Sites*, 2005, 7(2); UNESCO, *Cultural Heritage and Outstanding Universal Value*, Walnut Creek, CA: AltaMira Press, 2013; 'The Upstream Process: The Way Forward for the World Heritage Convention?', *Heritage & Society*, 2014, 7(1).

26 See *Cultural Tourism in India: Its Scope and Development With Special Reference to the Monumental Heritage*, by F. R. Allchin, May 1969. UNESCO archives, file reference FR/TA/CONSULTANT http://unesdoc.unesco.org/images/0000/000088/008820EB.pdf (accessed on 20 December 2016).

27 Ray, 'Introduction World Heritage Cultural Sites in India: Defining Parameters', in Kumar (ed.), *Indian World Heritage Sites in Context*, New Delhi: National Monuments Authority and Aryan Books International, 2014; 'From Multi-Religious Sites to Mono-Religious Monuments in South Asia: The Colonial Legacy of Heritage Management', in P. Daly and T. Winter (eds.), *The Routledge Handbook of Heritage in Asia*, London: Routledge, 2007.

28 Ray (ed.), *Mausam: Maritime Cultural Landscapes Across the Indian Ocean*, New Delhi: Aryan Books International, 2014.

29 M. Wheeler, *Still Digging*, London: Pan Press, 1958, pp. 156 and 161.

30 www.intach.org

31 A. G. Krishna Menon, 'Rethinking the Venice Charter: The Indian Experience', *South Asian Studies*, 1994, 10(1); 'Introduction to the Exhibition: Delhi: A Living Heritage', in *Delhi: A Living Heritage*, New Delhi: IGNCA and INTACH, 2010. For a critique of INTACH see A. Krishna, 'The Care and Management of Historic Hindu Temples in India: An Examination of Preservation Policies Influenced by the Venice Charter in Non-Judeo-Christian Contexts', *Change Over Time*, 2014, 4(2). She states that practitioners like A. G. Krishna Menon, Thakur and Prasad have offered alternative approaches to preservation in India that correspond to local practices.

32 T. Harkness and A. Sinha, 'Taj Heritage Corridor: Intersections Between History and Culture on the Yamuna Riverfront [Places/Projects]', *Places*, 2004, 16(2). See especially Krishna, 'The Care and Management of Historic Hindu Temples in India: An Examination of Preservation Policies Influenced by the Venice Charter in Non-Judeo-Christian Contexts'.

33 'The Care and Management of Historic Hindu Temples in India: An Examination of Preservation Policies Influenced by the Venice Charter in Non-Judeo-Christian Contexts'.

34 See for example E. Herbert, 'Curzon Nostalgia: Landscaping Historical Monuments in India', *Studies in the History of Gardens & Designed Landscapes: An International Quarterly*, 2012, 32(4); Krishna, 'The Care and Management of Historic Hindu Temples in India: An Examination of Preservation Policies Influenced by the Venice Charter in Non-Judeo-Christian Contexts'.

35 Aga Khan Development Network, 'Nizamuddin Urban Renewal Initiative 2015 Annual Report', AKTC, 2015.

36 S. Kavuri-Bauer, *Monumental Matters: The Power, Subjectivity, and Space of India's Mughal Architecture*, Durham, NC: Duke University Press, 2011; S. Prasad and K. Gavsker, 'The New "Love" Story of the Taj Mahal: Urban Planning in the Age of Heritage Tourism in Agra', *Review of Urban Affairs*, 2016, 51(5).

37 J. M. Fritz and G. Michell, 'Living Heritage at Risk: Searching for a New Approach to Development, Tourism, and Local Needs at the Grand Medieval City of Vijayanagara', *Archaeology*, November/December 2012; N. Bloch, 'Evicting Heritage: Spatial Cleansing and Cultural Legacy at the Hampi UNESCO Site in India', *Critical Asian Studies*, 2016, 48(4); 'Barbarians in India: Tourism as Moral Contamination', *Annals of Tourism Research*, 2017: 62.

38 Comptroller and the Auditor General of India, 'Report on Performance Audit of Preservation and Conservation of Monuments and Antiquities',

Report No. 19 of 2013, Indian Audit and Accounts Department, 2013. See the highlight of audit findings, pp. viii–xi, www.cag.gov.in/sites/default/files/audit_report_files/Union_Performance_Ministry_Cultures_Monuments_Antiquities_18_2013.pdf (accessed on 14 January 2017).

39 D. K. Chakrabarti, 'Archaeology and Politics in the Third World, With Special Reference to India', in R. Skeates, C. McDavid and J. Carman (eds.), *The Oxford Handbook of Public Archaeology*, Oxford: Oxford University Press, 2012.

40 Comptroller and the Auditor General of India, 'Report on Performance Audit of Preservation and Conservation of Monuments and Antiquities', p. 39.

41 C. Nakamura, 'Mumbai's Quiet Histories: Critical Intersections of the Urban Poor, Historical Struggles, and Heritage Spaces', *Journal of Social Archaeology*, 2014, 14(3): 272–273.

42 See http://whc.unesco.org/en/list/1338

43 S. Ahmed, 'Jantar Mantar, Jaipur: Implementing the Management Plan', *Context*, 2013, X(2).

44 S. Jain, 'Indian Heritage Cities Netowrk: Walking Into the Microcosm of Jaipur', Developed for the Minister for Urban Development, Housing, Local Self Government, Home, Law, Parliamentary Affairs, Government of Rajasthan, 2011. See also Y. Narayanan, *Religion, Heritage and the Sustainable City: Hinduism and Urbanisation in Jaipur*, London: Routledge, 2014.

45 See the organisation's website (www.dronah.org/about.aspx) for information on training workshops and other activities, heritage walks, internships and volunteers, and grants offered.

46 http://whc.unesco.org/en/list/232

47 Aga Khan Development Network, "Nizamuddin Urban Renewal Initiative 2015 Annual Report'. There has been co-funding from new partners, including the Sir Dorabji Tata Trust, Ford Foundation, World Monuments Fund, the US Ambassador's Fund for Cultural Preservation, the Embassy of the Federal Republic of Germany, Sir Ratan Tata Trust, the Delhi Urban Heritage Foundation, as well as the Municipal Corporation of Delhi.

48 Aga Khan Development Network, 'Nizamuddin Urban Renewal Initiative 2015 Annual Report'.

49 Aga Khan Development Network, 'Nizamuddin Urban Renewal Initiative 2015 Annual Report'.

50 http://whc.unesco.org/en/list/252. This mausoleum of white marble was built in Agra between 1631 and 1648 by order of the Mughal emperor, Shah Jahan, in memory of his favourite wife.

51 http://whc.unesco.org/en/list/251 Near the gardens of the Taj Mahal stands the 16th-century Mughal monument known as the Red Fort of Agra. The fortress built of red sandstone encompasses, within its 2.5-km-long enclosure walls, the imperial city of the Mughal rulers.

52 See the Partners page on the organisation's website (http://cureindia.org).

53 See D. Brockington, *Fortress Conservation: The Preservation of the Mkomazi Game Reserve, Tanzania*, Bloomington, IN: Indiana University Press, 2004; D. Byrne, 'Archaeology and the Fortress of Rationality', in L. M. Meskell (ed.), *Cosmopolitan Archaeologies*, Durham, NC: Duke University Press, 2009; Meskell, *The Nature of Heritage: The New South Africa*.

54 Nakamura, 'Mumbai's Quiet Histories: Critical Intersections of the Urban Poor, Historical Struggles, and Heritage Spaces', p. 290.

Bibliography

Aga Khan Development Network, *Nizamuddin Urban Renewal Initiative 2015 Annual Report*, New Delhi: AKTC, 2015.

Ahmed, S., 'Jantar Mantar, Jaipur: Implementing the Management Plan', *Context*, 2013, X(2): 141–148.

Bendix, R., 'The Power of Perseverance: Exploring the Negotiation Dynamics at the World Intellectual Property Organization', in B. Müller (ed.), *The Gloss of Harmony: The Politics of Policy-Making in Multilateral Organisations*, London: Pluto Press, 2013, pp. 23–45.

Bertacchini, E., C. Liuzza, and L. Meskell, 'Shifting the Balance of Power in the UNESCO World Heritage Committee: An Empirical Assessment', *International Journal of Cultural Policy*, 2017, 23(3): 331–351.

Bertacchini, E., C. Liuzza, L. M. Meskell, and D. Saccone, 'The Politicization of UNESCO World Heritage Decision Making', *Public Choice*, 2016, 167(1–2): 95–129.

Betts, P., 'Humanity's New Heritage: UNESCO and the Rewriting of World History', *Past & Present*, 2015, 228(1): 249–285.

Betts, P., 'The Warden of World Heritage: UNESCO and the Rescue of the Nubian Monuments', *Past & Present*, 2015, 226(suppl 10): 100–125.

Bloch, N., 'Evicting Heritage: Spatial Cleansing and Cultural Legacy at the Hampi UNESCO Site in India', *Critical Asian Studies*, 1 October 2016, 48(4): 556–578.

———, 'Barbarians in India: Tourism as Moral Contamination', *Annals of Tourism Research*, 2017, 62(1): 64–77.

Brockington, D., *Fortress Conservation: The Preservation of the Mkomazi Game Reserve, Tanzania*, Bloomington, IN: Indiana University Press, 2004.

Byrne, D., 'Archaeology and the Fortress of Rationality', in L. M. Meskell (ed.), *Cosmopolitan Archaeologies*, Durham, NC: Duke University Press, 2009, pp. 68–88.

Chakrabarti, D. K., 'Archaeology and Politics in the Third World, With Special Reference to India', in R. Skeates, C. McDavid and J. Carman (eds), *The Oxford Handbook of Public Archaeology*, Oxford: Oxford University Press, 2012, pp. 117–131.

Choay, F., *The Invention of the Historic Monument*, Cambridge: Cambridge University Press, 2001.

Claudi, I. B., 'The New Kids on the Block: BRICs in the World Heritage Committee', MA Thesis, Department of Political Science, University of Oslo, Norway, 2011.

Comptroller and the Auditor General of India, 'Report on Performance Audit of Preservation and Conservation of Monuments and Antiquities', Report No. 19 of 2013, Indian Audit and Accounts Department, 2013.

De Cesari, C., 'World Heritage and Mosaic Universalism', *Journal of Social Archaeology*, 2010, 10(3): 299–324.

De Cesari, C., 'World Heritage and the Nation-State', in Chiara De Cesari and Ann Rigney (eds), *Transnational Memory: Circulation, Articulation, Scales*, Berlin: de Gruyter, 2014, pp. 247–270.

Deese, R. S., *We Are Amphibians: Julian and Aldous Huxley on the Future of Our Species*, Berkeley, CA: University of California Press, 2014.

Fritz, J. M., and G. Michell, 'Living Heritage at Risk: Searching for a New Approach to Development, Tourism, and Local Needs at the Grand Medieval City of Vijayanagara', *Archaeology*, November/December 2012: 55–62.

Harkness, T., and A. Sinha, 'Taj Heritage Corridor: Intersections Between History and Culture on the Yamuna Riverfront [Places/Projects]', *Places*, 2004, 16(2): 62–69.

Herbert, E., 'Curzon Nostalgia: Landscaping Historical Monuments in India', *Studies in the History of Gardens & Designed Landscapes: An International Quarterly*, 2012, 32(4): 277–296.

Hoggart, R., *An Idea and Its Servants: UNESCO From Within*, Piscataway, NJ: Transaction, 2011.

Jain, S., 'Indian Heritage Cities Netowrk: Walking Into the Microcosm of Jaipur', Developed for the Minister for Urban Development, Housing, Local Self Government, Home, Law, Parliamentary Affairs, Government of Rajasthan, 2011.

Jain, S., and R. Hooja, 'The Hill Forts of Rajasthan as a Serial Nomination', *Context*, 2013, X(2): 77–87.

Joy, C., *The Politics of Heritage Management in Mali: From UNESCO to Djenné*, Walnut Creek, CA: Left Coast Press, Inc., 2012.

Joy, C., ' "UNESCO Is What?" World Heritage, Militant Islam and the Search for a Common Humanity in Mali', in C. Brumann and D. Berliner (eds), *World Heritage on the Ground: Ethnographic Perspectives*, Oxford: Berghan, 2016.

Kavuri-Bauer, S., *Monumental Matters: The Power, Subjectivity, and Space of India's Mughal Architecture*, Durham, NC: Duke University Press, 2011.

Krishna, A., 'The Care and Management of Historic Hindu Temples in India: An Examination of Preservation Policies Influenced by the Venice Charter in Non-Judeo-Christian Contexts', *Change Over Time*, 2014, 4(2): 358–386.

Krishna Menon, A. G., 'Rethinking the Venice Charter: The Indian Experience', *South Asian Studies*, 1994, 10(1): 37–44.

Krishna Menon, A. G., 'Introduction to the Exhibition: Delhi: A Living Heritage', in *Delhi: A Living Heritage*, New Delhi: IGNCA and INTACH, 2010, pp. 2–15.

Labadi, S., 'A Review of the Global Strategy for a Balanced, Representative and Credible World Heritage List 1994–2004', *Conservation and Management of Archaeological Sites*, 2005, 7(2): 89–102.

Labadi, S., *UNESCO, Cultural Heritage and Outstanding Universal Value*, Walnut Creek, CA: AltaMira Press, 2013.

Labadi, S., 'The Upstream Process: The Way Forward for the World Heritage Convention?', *Heritage & Society*, 2014, 7(1): 57–58.

Mazower, M., *No Enchanted Palace: The End of Empire and the Ideological Origins of the United Nations*, Princeton, NJ: Princeton University Press, 2009.

Mazower, M., 'The End of Eurocentrism', *Critical Inquiry*, 2014, 40(4): 298–313.

Meskell, L. M., 'Archaeological Ethnography: Conversations Around Kruger National Park', *Archaeologies: Journal of the World Archaeology Congress*, 2005, 1(1): 83–102.

Meskell, L. M., 'From Paris to Pontdrift: UNESCO Meetings, Mapungubwe and Mining', *South African Archaeological Bulletin*, 2011, 66(194): 149–156.

Meskell, L. M., *The Nature of Heritage: The New South Africa*, Oxford: Wiley Blackwell, 2011.

Meskell, L. M., 'The Rush to Inscribe: Reflections on the 35th Session of the World Heritage Committee, UNESCO Paris, 2011', *Journal of Field Archaeology*, 2012, 37(2): 145–151.

Meskell, L. M., 'States of Conservation: Protection, Politics and Pacting Within UNESCO's World Heritage Committee', *Anthropological Quarterly*, 2014, 87(1): 267–292.

Meskell, L. M., 'Gridlock: UNESCO, Global Conflict and Failed Ambitions', *World Archaeology*, 2015, 47(2): 225–238.Meskell, L. M., 'World Heritage and WikiLeaks: Territory, Trade and Temples on the Thai-Cambodian Border', *Current Anthropology*, 2016, 57(1): 72–95.

Meskell, L. M., C. Liuzza, E. Bertacchini, and D. Saccone, 'Multilateralism and UNESCO World Heritage: Decision-Making, States Parties and Political Processes', *International Journal of Heritage Studies*, 2015, 21(5): 423–440.

Müller, B., 'Lifting the Veil of Harmony: Anthropologists Approach International Organizations', in B. Müller (ed.), *The Gloss of Harmony: The Politics of Policy-Making in Multilateral Organisations*, London: Pluto Press, 2013, pp. 1–20.

Nakamura, C., 'Mumbai's Quiet Histories: Critical Intersections of the Urban Poor, Historical Struggles, and Heritage Spaces', *Journal of Social Archaeology*, 2014, 14(3): 271–295.

Narayanan, Y., *Religion, Heritage and the Sustainable City: Hinduism and Urbanisation in Jaipur*, London: Routledge, 2014.

O'Neill, J., *Building Better Global Economic BRICs*, New York: The Goldman Sachs Group, 2001.

Pavone, V., 'From Intergovernmental to Global: UNESCO's Response to Globalization', *The Review of International Organizations*, 2007, 2(1): 77–95.

Pavone, V., *From the Labyrinth of the World to the Paradise of the Heart: Science and Humanism in UNESCO's Approach to Globalization*, New York: Lexington, 2008.

Plets, G., 'Ethno-Nationalism, Asymmetric Federalism and Soviet Perceptions of the Past: (World) Heritage Activism in the Russian Federation', *Journal of Social Archaeology*, 2015, 15(1): 67–93.

Prasad, S., and K. Gavsker, 'The New "Love" Story of the Taj Mahal: Urban Planning in the Age of Heritage Tourism in Agra', *Review of Urban Affairs*, 2016, 51(5): 40–48.

Ray, H. P., 'From Multi-Religious Sites to Mono-Religious Monuments in South Asia: The Colonial Legacy of Heritage Management', in P. Daly and T. Winter (eds), *The Routledge Handbook of Heritage in Asia*, London: Routledge, 2007, pp. 69–84.

Ray, H. P., 'Introduction World Heritage Cultural Sites in India: Defining Parameters', in M. Kumar (ed.), *Indian World Heritage Sites in Context*. New Delhi: National Monuments Authority and Aryan Books International, 2014.

Ray, H. P. (ed.), *Mausam: Maritime Cultural Landscapes Across the Indian Ocean*, New Delhi: Aryan Books International, 2014.

Shore, C., 'European Integration in Anthropological Perspective: Studying the "Culture" of the EU Civil Service', in R. A. W. Rhodes, P. t' Hart and M. Noordegraaf (eds), *Observing Government Elites: Up Close and Personal*, New York: Palgrave Macmillan, 2007, pp. 180–205.

Sluga, G., 'UNESCO and the (One) World of Julian Huxley', *Journal of World History*, 2010, 21(3): 393–418.

Sluga, G., 'Editorial – the Transnational History of International Institutions', *Journal of Global History*, 2011, 6(2): 219–222.

Sluga, G., 'Imagining Internationalism', *Arts: The Journal of the Sydney University Arts Association*, 2012, 32: 55–68.

Sluga, G., *Internationalism in the Age of Nationalism*, Philadelphia, PA: University of Pennsylvania Press, 2013.

Toye, J., and R. Toye, 'One World, Two Cultures? Alfred Zimmern, Julian Huxley and the Ideological Origins of UNESCO', *History*, 2010, 95(319): 308–331.

Wheeler, M., *Still Digging*, London: Pan Press, 1958.

3 Valuation of world heritage

Indira Rajaraman

Heritage has been absent from the mainstream sustainable development debate despite its crucial importance to societies and the wide
acknowledgment of its great potential to contribute to social, economic
and environmental goals. World Heritage may provide a platform to
develop and test new approaches that demonstrate the relevance of
heritage for sustainable development, with a view to its integration in
the UN post-2015 development agenda.

— UNESCO World Heritage Centre[1]

Introduction

World Heritage Sites fall in two categories: cultural heritage and natural heritage.[2] Outstanding Universal Value – the basis for identification of a World Heritage Site – implies that valuation, from a global
perspective, might be sufficiently greater than the valuation of that
site in its immediate location for the World Heritage tag to provide a
significant incentive for the preservation of it. The rise in tourist traffic consequent upon inclusion[3] in the World Heritage List, statistically
validated in several contexts, seemingly justifies that prior.

The common cliché of the historical structure allowed to fall into
ruin, its stones prised out for constructing nearby dwellings, or the
area within what remains of the structure put to base uses, conforms
to that conception. Likewise, there are natural biodiversity reserves
threatened by unregulated cutting of trees where inclusion on the
World Heritage List may well be the statement of superior valuation
needed to put a stop to such practices and preserve what remains. This
chapter does not therefore argue that the World Heritage convention
should be abandoned, and its sites de-recognised. What it does argue
is that the basis of indigenous valuation should be investigated for its
constituents, so as to bring about a better alignment between global
and local valuation.

In recent years, as exemplified by the quote that begins this chapter, there has been an attempt to integrate heritage with sustainability, the new global watchword underlying post-2015 Sustainable Development Goals. In successive meetings of the World Heritage Committee going back to 1994, the operational guidelines increasingly pay attention to sustainable preservation of cultural and natural heritage. In the extended deliberations to mark the 40th anniversary of the World Heritage Convention in 2012, there was a Toyama proposal to mainstream heritage in sustainable policies, and sustainability in heritage policies and practices, but the proposal contains little beyond the need to manage the inevitable tension between preservation of biodiversity and forest wealth on the one hand and livelihoods of people in those environments on the other, and between tourism access to World Heritage Sites, and the destructive aspects of that access.

The 2011 amendment by the World Heritage Committee of the definition of sustainable development in the operational guidelines[4] does mention that World Heritage Sites must advance the quality of life of local communities, but the specifics of what that may entail are not spelled out. Concern is also expressed about how best to convey decisions that might be taken on sustainable management of heritage, and the implementation of these policies on the ground in what now number more than 1,000 heritage sites. There are so far only two examples of a listed site actually having been de-listed,[5] but there is a danger list which hosted two sites in India for a short period: one natural site (Manas Natural Sanctuary), and one cultural (Hampi).[6]

The next section will look at indigenous valuation, and what that exploration leads us to. The section following looks at the all-important issue of funding for preservation. The subsequent two sections look respectively at World Heritage Sites in India against the background of the preceding sections, and at possible expansion of the cultural criteria for World Heritage listing to include traditional knowledge systems for preservation of water, where the physical structures submitted for consideration to the World Heritage Committee would be distinguished by their ingenuity and functionality rather than any particular aesthetic quality. It is also argued that until World Heritage recognition extends to sites containing fossilised remains of dinosaurs and other even earlier pre-historic creatures – sites which have provided vital clues to how land masses and pre-historic life evolved over time – the conception of Outstanding Universal Value will remain seriously limited and deeply flawed. There is a final concluding section.

It is important to state upfront what this chapter does not do. The notion of development as an aspirational ideal is not questioned, nor is it taken as necessary that development calls for the abandonment of traditional culture and values.[7] Choosing development is emphatically not seen as equivalent to choosing a particular strategy of development such as that for instance advanced by the Washington Consensus. Finally, the chapter does not investigate the issue of whether World Heritage listing or classification might have mis-specified the categorisation or positioning of a site in a region's heritage.[8]

Indigenous valuation of a cultural or natural site

Valuation is dealt with in the economist's sense of preferences between two binary alternatives *ceteris paribus*, which may be single-option choices (for example between having enough water for survival or not having enough), or bundled options between cost–benefit packages. In single-option choices, the chapter assumes for simplicity that some choices can be made without reference to indigenous communities – in this example, having water rather than not having it. It does not touch upon what is by now a fairly large literature on how valuation can be quantified through survey-based assessments of contingent valuation (willingness to pay or rank orderings), or hedonic pricing models evaluating the impact on local real estate value of restoring World Heritage Sites.[9] The focus instead is on the basis of traditional valuation, prior to restoration, and on context-specific valuation rather than any macro sense of the contribution of heritage or culture to wellbeing.

By definition, a cultural site which exists in the present day has survived, which is to say that the valuation of it as revealed by its existence is sufficiently non-negative for it not to have been demolished. This could be single-point valuation, as for instance by a government which has effective jurisdiction over an area, and might have legislation marking protected sites. Or there might be polycentric valuation, where at the limiting extreme each individual living in the area could be one such polycentric valuation centre. In the latter case, the existence of the cultural site could be thought of as an equilibrium outcome with non-negative valuations trumping negative valuations. This equilibrium could well be suddenly overturned, as for example in the case of the Bamiyan Buddhas, which were blasted out in 2001 when their very existence was an affront to a powerful subset of individuals residing in the area.

Cultural monuments had to have been conceived and executed by powerful individuals (rulers or successful merchants) with access to large financial resources. Their motivations would have ranged from the usual projection of power and grandeur, to altruism at the other extreme – as in the case of forts or palaces constructed to provide public employment in times of famine. Whatever the motivation, the monument has to have had some intersection with the values of the local community for it to have been regarded with favour in its immediate vicinity, enough to ensure its survival in what was typically an unguarded state.

The value accorded by the local community to the site could be based on any or all of five bundles of rights ordered hierarchically (as conceptualised by Edella Schlager and Elinor Ostrom).[10] At the most basic level, there might be access rights and withdrawal rights. Access could be access to water sources enabled by the site, or just access to an open space, or the perceived spiritual advantages of proximity to a structure with religious or political significance.[11] The powers of benediction gained from an initial such site could then form the nucleus of multiple sites.[12]

Withdrawal could be traditionally assigned rights to renewable resources such as leaves, fruit, flowers or bark from the site (even cultural sites might have attached gardens or orchards), or even extraction of timber or firewood, in volumes that have been kept sustainable.

Three other bundles of rights at increasing levels of ordering to complete the Schlager–Ostrom set of five are management rights: the right to manage and regulate internal use; exclusion rights to decide who can exercise the rights of access, withdrawal or management; and alienation rights, including the right to lease or sell any of the first four rights.

Typically, this last set of management, exclusion and alienation rights, even if assigned at the time the monument was constructed, might have been lost over time to powerful predators, leading to the commonly observed phenomena of overuse or active abuse. Where access and withdrawal rights are highly valued, as for example, access to water enabled by the construction of the monument, simultaneously with the loss of management and exclusion rights, one would see typically dense human settlements around the monument, where the water source itself might get depleted through overuse, or polluted and unusable through mis-use.

This example, of water access bundled with a cultural site, demonstrates that a neat separation is not always possible in practice along the lines of the useful conceptual distinction between common-pool resources and cultural commons made by Enrico Bertacchini and his colleagues.[13] Both categories suggested by them carry the property of non-excludability (or low possibility of exclusion). But where common-pool resources share with private goods the attribute of rivalry, cultural commons are non-rival. Although cultural commons do not therefore suffer from limited carrying capacity, like common-pool, they suffer from the classic public good problem of free-riding and under-provision (or under-conservation, in the cultural context). Where cultural commons in the form of a monument is bundled with a common-pool resource like water, there is a particularly unfortunate combination of natural deterioration of the monument, coupled with depletion and pollution of the water source that goes with it.

This circumstance, of defiled and encroached sites, is paradoxically not an indicator that the site carries no indigenous local valuation. On the contrary, it indicates that the site carried (and still does) very high value to the local community, enough to attract (and retain) human settlement. These settlements might have developed over the years cultural and craft traditions specific to them, perhaps dating from the time of construction. The restoration of the monument which is done to build candidacy for World Heritage listing has to recognise the basis of indigenous valuation, so that the resulting protected monument is not built on a platform of alienation from the surrounding community. Enhanced tourism could create such a barrier between the site and the surrounding community, if the restored site has not succeeded in connecting with the local community by respecting what drew community members to the site in the first place. There has to be more than ensuring a local share in tourism revenue, or access to licenced service provision at the site, for a more fundamental alignment of global and local value leading to sustainable integration of the site with its environment.

Natural sites such as forests share the same fundamental preservation problems as cultural sites. They have traditionally carried access and withdrawal rights, extending even to residence rights within the preserve, which typically go with management rights, in terms of control of internal use. What often happens with powerful outside groups seeking to appropriate the timber value of forests is that the traditional forest residents lose control over exclusion rights, and with that the degradation of forest cover sets in. In such cases, restoring control over exclusion rights to local residents, with external oversight to guard against capture by elites among local residents, may hold the key to preservation of these sites.

The state of forests is not thought to be crucially a function of the general type of forest governance by scholars who have studied forest conditions:[14] rather, it is how a particular governance arrangement fits the local ecology, how specific rules are developed and adapted over time, and whether users consider the system to be legitimate and equitable. We come back once again to the issue of valuation by the local community as a key element in whether efforts at forest preservation succeed, either with World Heritage listing or independent of it.

It costs more to preserve traditional access and withdrawal rights, along with restoration of the monument or nature reserve. The funding issue for restoration therefore becomes paramount.

The funding issue

The World Heritage tag carries no financing promise in the sense of a formal arrangement either before or after listing. The small World Heritage Fund[15] is financed by compulsory contributions by member countries, supplemented by ad hoc donations from private parties and governments.[16] There is a provision for countries to seek assistance from the Fund, but the flows into and out of the Fund do not seem to be very transparent. For that matter, there seems to be increasing evidence[17] that the very process of selection for inscription of World Heritage Sites is becoming increasingly opaque and unrelated to the recommendations of UNESCO's Advisory Bodies. In the case of India, in particular, the budgetary evidence suggests that India is a net contributor to the Fund, not a net recipient (at least in recent years).[18]

Even so, the prestige of inclusion in the World Heritage List is so great that large sums of money are expended in the process of applying to the World Heritage Committee. Since the value of World Heritage listing can be preserved only if the list is controlled, there is a high rejection percentage – and its magnitude does not seem to be widely known.

The insignificant provision for global funding is in itself indicative of the business model underlying World Heritage listing: recovery of costs borne by national or subnational governments on restoration through tourism revenue, over time. The World Heritage tag bestows commercial value by virtue of the enhanced tourist traffic that goes with the tag (for an econometric validation of this impact, see Brian Logan VanBlarcom and Cevat Kayahan[19]), and so raises the post-listing local value of the structure. However, case studies of cultural sites in the Western world show that even with enhanced traffic, tourism revenue covers only operating costs – and even that, never fully. Capital costs borne by national or subnational governments are never recovered.[20]

The balance between costs and incremental tourist revenue may be more favourable in contexts like India, where wages of skilled craftspeople are low. So, for countries such as India the business model of the World Heritage tag may carry greater validity in terms of cost cover than in developed countries, but only if the cost of restoration is narrowly confined to a ring-fenced site, without attention to the local community and its relationship to the pre-restoration site. Clearly, the present business model disincentivises any restoration costs extraneous to the bid to make the site a source of tourist revenue – and that is exactly what happens, alienating the surrounding population by destroying the place the site has in their scheme of things, and violating the sustainability concern expressed by the World Heritage Committee.

There are additional problems in India with the way in which tourism receipts are fiscally structured. If the tourism revenue from a site is an increasing function of the level of maintenance, simple incentive compatibility requires that the revenue collected at a site be sequestered at the site. But formal accounting arrangements in India do not permit the tourism revenue from an archaeological site to accrue to the site. The Archaeological Survey of India (ASI), which is charged with maintaining World Heritage Sites, gets an annual budgetary expenditure allocation from the Government of India, unrelated to tourism receipts which go into the general non-tax revenue pool.[21] Site independence and fiscal autonomy are necessary elements in any accountability mechanism. At the same time, it is also necessary that ASI be reformed and more effectively monitored so that is accountable for the budgetary allocation to it. The state of ASI monuments has been severely criticised by the Comptroller and Auditor General (CAG), the national auditor.[22]

In order for World Heritage listing to preserve indigenous valuation of the pre-restoration site, World Heritage Sites have to be restored in partnership with restoration of local surrounding human settlements, such that global and local valuation are fully aligned. Assigning surrounding communities formal space in the precincts where they can demonstrate their craft traditions, dating possibly from the time of construction of the monument or earlier, will ensure their co-operation in the restoration process and subsequent upkeep. But these and other ways of including the surrounding population in the restored site will clearly call for more funding than a ring-fenced site. Traditional funding patterns for sites highly valued in their vicinity for their religious or spiritual worth may no longer be possible to resurrect.

Local government, among whose functions is the provision of local public goods such as sanitation and road access, is a rightful source from which to seek fiscal assistance for the upgrade of surrounding human settlements. These funding partnerships are legitimate and necessary, but will require sharing of tourist revenue with them as well. Funding partnerships from commercial entities like hotels, which could gain from a more widely defined notion of site preservation, resulting in visitors actually spending time in the vicinity of a site rather than just narrowly targeting a monument, need to be explored. Leonardo Becchetti and his colleagues propose a earmarked tourist tax along Pigouvian lines, levied both on visitors to the site and goods and services sold in its larger vicinity, some of which can be retained for valorisation at the jurisdiction of collection, the rest pooled and redistributed on a regional scale.[23]

There have been some other very interesting suggestions in the recent literature on possible funding avenues for preservation of heritage. Corporate social responsibility (CSR) is one such. Becchetti and his colleagues[24] cite empirical evidence affirming that CSR is not necessarily negatively related to corporate profit performance. Their game-theoretic model suggests that selling a bundled product with cultural responsibility might be superior to an unbundled alternative.

Bruno Frey and Paolo Pamini[25] propose a scheme whereby each site on a globally agreed list will be assigned World Heritage Units (WHU) based on its cultural importance and the distance to what may be best described as adequate restoration (or prevention of deterioration). Each country then pledges its commitment in terms of WHU to be restored. Every WHU funded earns the funding nation (or private entity) a World Heritage certificate, after assessment by a certification board. Since the price per WHU will vary across sites, the certificates will be tradeable on a global platform. Quite aside from the demands of this Coasian scheme – among other things, an unimpeachable certification process – the funding proposed remains country-based, which is an inherent limitation. Country-based funding is necessarily paid for by the taxpayer base of each country, to whom the benefits from the existence of globally valued heritage properties accrue very unevenly.

A more justifiable taxable base for maintenance of global public goods would be owners of internationally mobile factors of production (labour, capital), for whom the very idea of global public goods is more meaningful. Such a globally sourced fund to supplement local and national costs of preservation, would be more just in conception.

In the Indian federal structure, there is a within-country parallel of fiscal support from higher levels. There is a national value attached to forest preservation, resulting from the environmental externalities of standing forests, which is at odds with subnational states, for whom forests carry an opportunity cost in terms of lost revenue from timber, and conversion of forest land to industrial uses. In recent years therefore, there have been national grants going to states to compensate them for the economic burden of keeping their forest cover protected.

Finance Commissions in India are set up every five years to set states' shares of the tax revenues of the national government. The Thirteenth Finance Commission designed a forest grant over and above the tax share, to bridge the gap between national and state-level valuation, in the form of financial compensation for the opportunity cost of land preserved as forest and not converted to non-forest uses. Designed as a two-part absolute grant to states over its horizon 2010–2015, it was pro-rated to forest cover (see Appendix 3).[26] The Thirteenth Finance Commission also had a non-formulaic grant that covered specific state needs, which included preservation of cultural and natural heritage. However, this second avenue of heritage funding did not really provide a global template since it was not formulaic.[27]

Routine funding under a global provision, with formulaic allocation, is an imperative for heritage preservation going into the future. How might such a global fund be raised without country contributions? The levy of a global Tobin tax on financial transactions has been suggested as a possible global fund for financing climate change mitigation and adaptation.[28] It is yet to find acceptance – but without some such initiative, and the extension of its proceeds towards heritage preservation, the only sites successfully preserved will be those where funding partnerships with private donors have been forged.

World Heritage Sites in India

Out of 1,052 World Heritage Sites in the world, there are 36 in India, 28 cultural, 7 natural, and 1 mixed, listed in Appendix 2 along with the year of elevation to World Heritage status. Applications for 45 more are pending for consideration by the World Heritage Committee.

Indigenous valuation of these sites will be a function of whether a site is cultural or natural, and within cultural, the type of site.

Taking first the cultural sites, these are grouped in the table under four sub-categories (not formal sub-categories of the World Heritage system): caves; large sites, multiple structures; single monuments; and cultural currently functioning. The basis of indigenous valuation will vary between these.

Caves, of which there are four recognised World Heritage Sites, are of their nature removed from human settlements. The positive gain is that they are not easily accessible for possible defilement, but by the same token there are no surrounding residents to protect them from external predators intent upon destruction. For this category, World Heritage listing could be wholly positive in its outcomes, in the sense of imposing more active efforts at preservation than what might have been the case previously with, as always, the dangers posed by the enhanced tourist inflows consequent upon inclusion in the list.

Large archaeological sites were most often capital cities of erstwhile kingdoms, and fell into disuse owing to the shifting of the capital by the same or successor kingdoms. In most cases, as in the case also of single monuments, the very construction of these imposing structures could only have been possible with access to water either from a pre-existing source, or from a newly enabled network of water channels connected to surface or underground collection points. Since access to water is what from time immemorial has enabled human settlement, it is entirely natural that the water brought in for construction, and for the sustenance of the activity for which the site was constructed, would attract human settlements on its periphery. It is also the fundamental reason why the immediate neighbourhood values the site, as the enabler of livelihoods by virtue of access to water, in addition to any other reasons specific to the site.[29] These other reasons may be their religious value, or political pride in the power and sway of the kingdom which constructed it, or in the case of structures constructed (or extended) as an employment generation scheme during drought or famine, memory of what the structure signifies.

It should not be surprising therefore that large sites and single monuments are surrounded by areas with often quite dense human settlement. Indeed, the genesis of the move to declare them as sites of global value originates from the fear that these satellite settlements will defile the monument beyond redemption. However, the paradox is that density of settlement signals the basis of valuation of the site to its immediate neighbourhood, and is the very indicator that the basis of valuation still survives.

Fatehpur Sikri, the capital constructed by Akbar, was abandoned for reasons of both water scarcity and political turbulence, and is not surrounded by dense habitation for the very reason of its abandonment. That is the central paradox: the worse the congestion surrounding a monument, the higher the local valuation of the site.

A single-monument analogue of Fatehpur Sikri is Rani-ki-Vav in Patan, Gujarat. Here is a structure built over a stepwell (*vav*), a traditional stepped well capturing subterranean potable water. Because the location of these stepwells was dictated by topology and subterranean water aquifers, they became communal water sources publicly accessible to all, unconstrained by caste or other considerations, and so had important cultural implications as well. The selection of this as a World Heritage Site has had to do with the beauty of the structure surrounding the steps leading into the water. The World Heritage inscription refers to the geotectonic shift in the 13th century which swept away the Saraswati river feeding into the aquifer sources of the well, but the loss of the water took with it the key cultural significance of the site. As a monument to celebrate a source of water in a water-scarce region, it might at one time have been surrounded by human settlements, which drifted away when the well ran dry. The monument itself remained well-preserved in the desert. Here again is an example where the state of preservation is no indicator of its local value.

In cases where a site is surrounded by pressing human settlements, and is perhaps for that reason in a poorer state of preservation, listing of the monument alone while neglecting the water source which was the aspect of the site of enduring value to the local population is to negate local value and therefore to drive a wedge between global and local value.

A positive example of enhancement of local value is the restoration of Humayun's Tomb by the Aga Khan Foundation, which was part of a much larger project involving renewal of a neighbouring urban slum.[30] Here is a documented instance where the water source near the site, a stepwell, has been rendered functional as part of the restoration effort. The source of local value was made even more valuable, the settlement response to it rationalised, as part of the effort to restore a globally valued monument to its rightful grandeur.

Knowledge and pre-historic heritage

Recognition of knowledge systems which have sustained life, foremost among which is preservation of traditional water systems, will mainstream sustainability into heritage preservation. These knowledge systems may well have fallen into disuse, but the more predatory systems which have supplanted them are in danger of running dry. Resurrection of older systems whose hallmark was sustainability, from such documentation as might be available, or constructed from oral memory of people now living, is an urgent necessity. This process will be greatly enhanced if there is a knowledge plank in the World Heritage Convention, lending world recognition to traditional knowledge as a key component of human heritage. From a climate change perspective, this is now no longer an option so much as an imperative.

In the very nature of these systems, they are likely to have been in use in many parts of the world characterised by water scarcity, and would by the same token be replicable, so that listing as World Knowledge Heritage is not merely a location-specific preservation of a monument or natural resource, but a live body of survival techniques applicable to other regions of the world similarly threatened by water scarcity. There is evidence that these knowledge systems were widely transmitted in the ancient world.

Recognition of World Knowledge Heritage qualifies on the grounds of both mitigation and adaptation, if activation of these systems forestalls further predatory use of groundwater, or preferably even reverses the damage that has already occurred. But recognition under this provision can happen only after establishing that the knowledge system can and has been reactivated, and that it has resulted in restoring water and enabling life.

What of other knowledge systems, such as traditional medicine? Pre-modern cures and prophylactics have over the years been authenticated and co-opted into modern medicine. Quinine as a prophylactic and cure for malaria was one such. Other systems such as acupuncture co-exist with modern medical systems and do not need elevation to World Knowledge Heritage to get authenticated.

Preservation of water is in a different category. It is characterised by networks, which can get de-activated and ploughed into by a market in water that benefits a few at the expense of the collectivity that live in a region. It is this aspect of water as a collective resource, whose network features can be destroyed by agents outside the network, that calls for external recognition and protection. In a future in which climate change will threaten water supplies in many parts of the world, traditional water preservation systems clearly carry the property of Outstanding Universal Value which is the basis for recognition under the World Heritage Convention.

Fortunately, the six criteria currently listed for recognition of cultural sites already include one (v) that could provide the basis for such recognition as it stands:

> v) to be an outstanding example of a traditional human settlement, land-use, or sea-use which is representative of a culture (or cultures), or human interaction with the environment especially when it has become vulnerable under the impact of irreversible change;[31]

There already exists a water system recognised under this criterion, in Iran. A WH site there, the Shushtar Historic Hydraulic System, was listed in 2009 – a system of water mills, dams, tunnels and canals constituting an irrigation system. Another Iranian WH site included in 2015, the Cultural Landscape of Maymand, preserves the landscape of a self-sufficient nomadic pastoral community, recognised for its water collection and preservation systems, and for its three-phase migration over the year between summer settlements with seasonal buildings newly constructed each season, and permanent cave dwellings for the winter.

It is clear from these examples that World Heritage thinking has already moved a considerable distance towards recognising cultural knowledge heritage which is not manifested in large or aesthetically appealing structures (see also Charlotte Hess[32] and Aldo Buzio and Alessio Re[33] for the evolution of thinking regarding World Heritage listing). The integration of sustainability into heritage will be rendered complete only if physical structures, above or below ground, can be recognised with reference to their functionality in enabling human survival rather than with reference to their aesthetic appeal.[34]

Traditional systems of water networks in India as listed by the Centre for Science and Environment are prevalent in all parts of India, some names indicating the particular ruler in whose reign they came into existence.[35] These can be categorised under three broad types taken in turn. What distinguishes each is the particular combination of land topology and water source which enabled it. Not all may be fully or even partially functional today.

The first type was for sourcing drinking water. There are many simple rainwater catchment tanks, known by different regional names; the preservation/resuscitation of these is vital for the population of these regions, but they should be held to qualify for the Water Knowledge Heritage recognition only when they carry some special feature, such as networks of collection points linked to underground aquifers, or gravitational percolation in mountainous terrain. The idea is to recognise the engineering ingenuity in traditional water networks which enabled sustainable life in inhospitable terrain, and whose abandonment on account of predatory tapping of groundwater has led to distress in these regions, and forced outmigration in many cases. By the approach argued for in this chapter, if a defunct network can be activated through application of a traditional knowledge system, and can actually become a source of potable water over a wide area, connectable to surrounding human habitations with modern methods of access, such a resuscitated water network could then apply for listing as a World Heritage system, where its claim is based on preservation of traditional knowledge towards current usability, and usage.

A second type of water system is for harvesting of excess floodwater from full rivers during the monsoon, an example being the *ahar/pyne* system in Bihar. The network was actively dismantled during the British colonial regime, and the lack of any subsequent effort to reactivate it has subjected Bihar to repeated and destructive floods in the absence of the old catchment structures. The *ahar* was the catchment basin where monsoon overflow was retained, for use towards a second crop season, and *pynes* were channels constructed to lead the river water into the *ahar*.[36] Variants of this system for preservation of superfluous river water after the monsoon can be found all over the country, under different names, and with patterns specific to the topology and soil of each region. What would make any of these systems qualify for World Heritage recognition would be the restoration of the entire network in the region where they were once operative, and the documentation of their functionality and sustainable usability from one monsoon to the next.[37]

A third type of water preservation has to do not with preserving river overflow, but with harvesting rainwater on farmland with no supplementary sources of irrigation, such that the water saturates the land and enables successful sowing and harvesting. An example were *khadins/dhoras*, structures constructed in the 15th century in Jaisalmer, an arid region in west Rajasthan, said to be similar to methods used in ancient Ur in 4500 BCE and later in the Negev desert, and more recently by Native Americans in southwestern Colorado. *Surangams* for capture of rock seepage in rocky terrain are still functional to some degree, and are said to be similar to *qanats* which once existed in Mesopotamia and Babylon around 700 BCE.

Starting in 1984, there has been a move to activate 3,000 *johads*, networks of traditional earthen checkdams, across more than 650 villages in Alwar district, Rajasthan. The general rise of the groundwater level by almost 6 metres and in the forest cover in the area by 33% was documented in Anil Agarwal's classic 1999 work titled *Dying Wisdom*.[38] Five rivers that once went dry immediately following the monsoon have now become perennial.

Although the water harvesting systems discussed have cited only rural examples, there are a number of catchment networks in urban centres that have been subsumed and severely disrupted by unregulated real estate development. Recent episodes of severe monsoon season flooding in major urban centres like Mumbai (2012) and Chennai (2015), so severe that daily life was brought to a halt for a period of a week or more, are the result of blockage of natural rainwater runoff into catchment bodies, simultaneously causing the drying up of those water bodies and flooding the city. In these cases, the restoration of the original rainwater flow channels – which might call for reversal of some urban developments that have taken place and resettlement of those affected – call for another possible category of World Heritage recognition.

Finally, and perhaps most fundamentally, the very notion of Outstanding Universal Value has to be expanded to include pre-historic sites, which have provided such very essential clues to the evolution of the planet on which we live.[39] These sites are truly more global than the present very narrow conception of World Heritage provides for, in that they tell us the story of the evolution of life on earth, before nation states, before human life itself, indeed before the formation of the present land mass on earth as we know it.

Until the boundaries of UNESCO recognition are expanded to include these, the most truly global of all sites for the value of the story they tell, the conception of World Heritage will remain limited and flawed. These sites are seriously threatened, since national archaeological bodies are confined in their conception to the evolution of human culture within their national jurisdictions. In some cases, they do include within their list of protected sites caves with rock paintings from early human settlements (such as the Bhimbetka Rock Shelters in India which is also a World Heritage site, see Appendix 2), but much earlier sites which have provided vital clues to how land masses evolved over time are found in fields of fossilised remains of dinosaurs and other pre-historic creatures. These sites lie unprotected, with only a few committed individuals struggling with the judicial system in their respective countries to seek a stay of land-grabbing commercial interests. Here is an instance of how an unequivocal assertion of global value could rescue from obliteration the marks of the early years of our planet, which carry no immediate value in their immediate vicinity.

All these expansions of the boundaries of heritage recognition will be feasible only if global funding is made larger, and more transparent, than it presently is.

Summary

The business model underlying the World Heritage scheme is country funding of restoration against the promise of enhanced commercial value from higher tourist traffic consequent upon inclusion in the World Heritage List. The model disincentivises any restoration costs extraneous to the bid to make the site a source of tourist revenue. Cost minimisation requires that the site be ring-fenced from its immediate surroundings, at odds with access or withdrawal rights that may have been the traditional bases for local valuation, and running the risk therefore of local disengagement or worse, local hostility. The very density of human settlements around some of these sites is indicative of features such as water availability or other advantages of prox-imity which drew them, which might carry either rival or non-rival attributes. As for natural heritage sites such as forests, World Heritage listing without building on such access and withdrawal rights as made for (positive) local engagement with the site is not sustainable. It may actually serve to lower rather than raise the defences against external poaching and defilement.

The key issue is funding. There are many suggestions in the litera-ture reviewed in the chapter including Pigouvian tourist taxes pooled and redistributed within a region, Coasian schemes and inclusion of heritage in permissible directions for corporate social responsibility. The chapter mentions a successful example of a World Heritage Site in Delhi (Humayun's Tomb) where preservation has been extended to urban settlements adjoining the site itself, but since this was funded by private philanthropy, it is not widely replicable. Local government funding is a legitimate source to reach for in restoration of the wider area including human settlements surrounding a site, since many of the improvements call for attention to sanitation, road access and water – all local government functions. However, local governments are typi-cally weaker fiscally than higher levels of government. Other funding partnerships are possible with hotels and other such service providers who would gain commercially from a wider approach to restoration.

A more systemic solution calls for assistance from a global fund towards capital costs of restoration. The World Heritage Fund as it functions today is small, and access to it is not objective or formu-laic. Thus the first structural reforms needed are to raise the size of the World Heritage Fund and to make its fund allocation mechanisms more transparent, if World Heritage preservation is to better serve the objective of sustainability.

The chapter provides an example of within country funding in India through formula for forest preservation. Allocation of a global fund between World Heritage Sites by formula, whether for cultural or natural sites, will provide the seed funding needed to supplement the basic tourism based model. The best source for enhancement of the global fund is a Tobin tax levied at a small rate on international capi-tal flows. This shifts the base of tax collection to international capital flows, independent of country of origin, and therefore to a base more immediately linked to beneficiaries from globalisation.

The World Heritage scheme also needs to go beyond its present perimeters towards recognition of traditional knowledge systems which enabled life in ecologically challenging environments, and whose neglect poses grave problems of survival in the modern world. The key to identifying a traditional knowledge system has to be authenticity in resuscitation, and functionality in use in the modern world. Clause v of the ten criteria presently in place permits such a possible expansion in scope.

The chapter lists several indigenous water preservation networks in India characterised by ingenuity and effectiveness, documented by the Centre for Science and Environment, which fell into disuse because of predatory external forces. Revival of these is critically necessary in the context of climate change, towards both mitigation and adaptation, so that inclusion within the scope of World Heritage recognition will build sustainability into the very fabric of heritage preservation.

Finally, and equally urgently, the perimeters of World Heritage recognition need to be expanded to include fields of fossilised remains of dinosaurs and other even earlier pre-historic creatures – sites which have provided vital clues to how land masses and pre-historic life evolved over time. Until the boundaries of UNESCO recognition are expanded to include these, the most truly global of all sites for the value of the story they tell, the conception of World Heritage will remain seriously limited and deeply flawed.

Acknowledgements

The author thanks Himanshu Prabha Ray, who motivated the chapter and provided numerous leads, and Ratish Nanda, Lynn Meskell and the Centre for Science and Environment for very useful discussions, with the usual disclaimer. An earlier version of the chapter was presented at a conference on Heritage in Context: Balancing the Global with the Local, New Delhi, 20–22 August 2016.

Notes

1 UNESCO, 'Sustainable Development', http://whc.unesco.org/en/sustainable development/ (accessed on 9 May 2018).
2 The distinction between the two may be somewhat artificial, as pointed out in Pia Brancaccio, 'World Heritage Sites of Ajanta, Ellora and Elephanta', paper presented at conference on Heritage in Context: Balancing the Global With the Local, New Delhi, 20–22 August 2016.
3 The technical term used for inclusion in the World Heritage list is "inscription".
4 UNESCO, *Operational Guidelines for the Implementation of the World Heritage Convention*, Paris: UNESCO, 2011, para 119.

5 Enrico Bertacchini, Donatella Saccone and Walter Santagata, 'Embracing Diversity, Correcting Inequalities: Towards New Global Governance for the UNESCO World Heritage', *International Journal of Cultural Policy*, 2011, 17(3): 279.

6 In 2011, 31 properties worldwide were on the endangered list; see Bertacchini et al., 'Embracing Diversity', p. 279.

7 For a very effective survey of these ideas, see Sophia Labadi and Peter G. Gould, 'Sustainable Development: Heritage, Community, Economics', in Lynn Meskell (ed.), *Global Heritage: A Reader*, 1st edition, New York: John Wiley, 2015, pp. 196–216. Other difficult issues are also not entered into in the paper, such as whether traditional values should necessarily be upheld, in cases where they violate internationally accepted norms.

8 See Himanshu Prabha Ray, 'From Multi-Religious Sites to Mono-Religious Monuments in South Asia: The Colonial Legacy of Heritage Management', in Patrick Daly and Tim Winter (eds.), *Routledge Handbook of Heritage in Asia*, London and New York: Routledge, 2012, pp. 69–84.

9 For a comprehensive survey of these approaches, see D. O'Brien, *Measuring the Value of Culture*, London: Department for Culture, Media and Sport, 2010, pp. 22–33.

10 Edella Schlager and Elinor Ostrom, 'Property Rights Regimes and Natural Resources: A Conceptual Analysis', *Land Economics*, 1992, 68(3): 249–262. See also Elinor Ostrom, 'Beyond Markets and States: Polycentric Governance of Complex Economic Systems', Nobel Prize Lecture, Stockholm, 8 December 2009.

11 A persuasive treatise on how the colonial vision of Asian structures shifted focus from their essentially spiritual aspect, in a generic rather than faith-specific manner, to their aesthetic and artistic worth, can be found in Himanshu Prabha Ray, 'From Multi-Religious Sites to Mono-Religious Monuments in South Asia', pp. 69–84.

12 The Humayun's Tomb complex is said to have been developed laterally in proximity to the burial ground of Hazrat Nizamuddin, a revered Muslim saint, for this reason.

13 Enrico Bertacchini, Giangiacomo Bravo, Massimo Marrelli and Walter Santagala, 'Defining Cultural Commons', in E. Bertacchini, G. Bravo, M. Marrelli and W. Santagata (eds.), *Cultural Commons: A New Perspective on the Production and Evolution of Cultures*, Cheltenham: Edward Elgar Publishing, 2012, pp. 3–18.

14 Tanya Hayes, 'Parks, People, and Forest Protection: An Institutional Assessment of the Effectiveness of Protected Areas', *World Development*, 2006, 34(12): 2064–2075.

15 For the biennium 2014–2015 (inclusive of both calendar years), the World Heritage Fund amounted to $5.33 million; see UNESCO, *World Heritage Committee Fortieth Session, Istanbul, Turkey, 10–20 July 2016*, Paris: WHC/16/40.COM/15, 10 June 2016. The Indian contribution towards this, paid in September 2015, is listed as $48,630. The author is grateful to Claudia Liuzza who alerted her to this source.

16 For example, India is listed as having made an ad hoc contribution of $90,864 during September to November 2011 towards requests for international assistance which could not be met through the routine budget; see UNESCO, *World Heritage Committee Fortieth Session, Istanbul, Turkey, 10–20 July 2016*.

17 Lynn Meskell, Claudia Liuzza, Enrico Bertacchini and Donatella Saccone, 'Multilateralism and UNESCO World Heritage: Decision-Making, States Parties and Political Processes', *International Journal of Heritage Studies*, 2015, 21(5): 423–440. For an excellent data based analysis of decisions at each stage of the inscription process, see Enrico Bertacchini, Claudia Liuzza, Lynn Meskell and Donatella Saccone, 'The Politicisation of UNESCO World Heritage Decision Making', *Public Choice*, 2016, 167(1): 95–129. For a case study illustrating how World Heritage inscription can get entangled with negotiations on commercial contracts and military alignments, see Lynn Meskell, 'World Heritage and WikiLeaks: Territory, Trade and Temples on the Thai-Cambodian Border', *Current Anthropology*, 2016, 57(1): 72–95.

18 Obligatory contributions to the United Nations and other international organisations are nested under Budget Head 2061 (budgeted at 3.86 billion rupees in 2015–2016), and include expenditures on external affairs including diplomatic missions abroad. The constituents of this are not routinely reported, but from UNESCO WHC/16/40.COM/15 (see endnote 14), the assessed contribution of India, paid in September 2015, works out to approximately 3 million rupees. The receipts from external assistance of the Archaeological Survey of India, which is charged with maintaining WH sites, shows zero receipts in the last three years, towards a total annual expenditure of 6.58 billion rupees in 2015–2016.

19 Brian Logan VanBlarcom and Cevat Kayahan, 'Assessing the Economic Impact of a UNESCO World Heritage Designation', *Journal of Heritage Tourism*, 2011, 6(2): 143–164. For a review of a contrarian finding by Renee Prud'homme, see Bertacchini et al., 'Embracing Diversity', p. 286.

20 The Val di Cornia archaeological park in Italy (not a World Heritage site) is shown to have succeeded as it did because it operated as a commercial entity jointly owned by surrounding municipalities which fully capture tourism revenue (Labadi and Gould, 'Sustainable Development'). Even so, it did not manage to recover operating costs fully; the capital costs were covered by grants from the European Union, and the Italian government.

21 Receipts under Budget Head 0202 cover education, sports, art and culture, and lump together tuition fees with entry fees at museums and all sites maintained by the Archaeological Survey of India. The total collection was budgeted at 1.32 billion rupees in 2015–2016.

22 Comptroller and Auditor General of India (CAG), 'Performance Audit of Preservation and Conservation of Monuments and Antiquities of Union Government', Report No. 18, Ministry of Culture, 2013.

23 Bertacchini et al., 'Embracing Diversity', 2012, pp. 278–288. For a broader treatment of earmarked taxes in multiple contexts, see M. Wilkinson, 'Paying for Public Spending: Is There a Role for Earmarked Taxes?', *Fiscal Studies*, 1994, 15(4): 119–135.

24 Leonardo Becchetti, Nazaria Solferino and M. Elisabetta Tessitore, 'How to Safeguard World Heritage Sites? A Theoretical Model of "Cultural Responsibility"', University of Rome Centre for Economic and International Studies Tor Vergata Research Paper Series, 12: 6; No. 318, July 2014.

25 Bruno Frey and Paolo Pamini, 'Making World Heritage Truly Global: The Culture Certificate Scheme', University of Zurich Institute for Empirical Research in Economics Working Paper Series, ISSN 1424-0459, No. 419, June 2009.

26 'Report of the Thirteenth Finance Commission', December 2009.

27 Details on the heritage sites funded over 2010–2015 under this provision are available in Indira Rajaraman, 'Valuation of World Heritage', India International Centre Occasional Paper Series, No. 17, New Delhi, December 2016.

28 Indira Rajaraman, 'Climate Change and Other Wishes', *Mint*, 8 January 2016.

29 There might also be withdrawal rights, in respect of fruit or other produce from trees found even within cultural sites. These might also be abused, as when the trees themselves have been cut down for firewood or for use as timber.

30 The project website is www.nizamuddinrenewal.org (accessed on 18 February 2017). Another instance of stepwell restoration is on the grounds of the Qutb Shahi tombs (not a World Heritage site) in Telengana state (Yunus Lasania, 'Badi Baoli Provides a Silver Lining', *The Hindu*, 6 May 2016).

31 UNESCO, 'The Criteria for Selection', https://whc.unesco.org/en/criteria/ (accessed on 26 June 2018).

32 Charlotte Hess, 'Constructing a New Research Agenda for Cultural Commons', in E. Bertacchini et al. (eds.), *Cultural Commons: A New Perspective on the Production and Evolution of Cultures*, Cheltenham: Edward Elgar Publishing, 2012, pp. 19–35.

33 Aldo Buzio and Alessio Re, 'Cultural Commons and New Concepts Behind the Recognition and Management of UNESCO World Heritage Sites', in E. Bertacchini et al. (eds.), *Cultural Commons: A New Perspective on the Production and Evolution of Cultures*, Cheltenham: Edward Elgar Publishing, 2012, pp. 178–193.

34 Culture is defined in the economics literature as: 'Those customary beliefs and values that ethnic, religious and social groups transmit fairly unchanged from generation to generation' (Luigi Guiso, Paola Sapienze and Luigi Zingales, 'Does Culture Affect Economic Outcomes', *Journal of Economic Perspectives*, 2006, 20(2): 23–48. The key is inter-generational transmission, updated slowly through the generations. There is no necessary link to physical sites of aesthetic beauty, although that adds to the probability of recognition of World Heritage sites. See also Ray, 'From Multi-Religious Sites to Mono-Religious Monuments in South Asia' for the colonial view of cultural monuments which shifted focus to their aesthetic worth from the spiritual inheritance they represented.

35 Center for Science and Environment, 'Traditional Rural Rainwater Harvesting', http://www.rainwaterharvesting.org/Rural/Traditional.htm (accessed on 26 June 2018)

36 One particular instance of the resurrection of this system in the village of Dihra, 28 kilometres south of Patna city, between 1995 and 2000, has been documented; see Centre for Science and Environment, *State of the Environment Report: Traditional Water Systems*, 4th edition, New Delhi: Centre for Science and Environment, 2003. In the case of some though not all of these systems, their structure is documented, as for example Bengal's Inundation Channels by Sir William Willcocks, an irrigation expert who had prior experience in Egypt and Iraq.

37 A press report on one such activation of a Mauryan *ahar/pyne* system, through documentation extracted from old books and scriptures, is in Subhash Pathak, 'Mauryan Irrigation Revives Dry Magadh', *Hindustan Times*, 4 January 2018.
38 Anil Agarwal, *Dying Wisdom*, New Delhi: Centre for Science and Environment, 1999.
39 See Pranay Lal, *India: A Deep Natural History of the Indian Subcontinent*, New Delhi: Random House India, 2016.

Bibliography

Aga Khan Development Network, 'Nizamuddin Urban Renewal Initiative', www.nizamuddinrenewal.org (accessed 26 June 2018).

Agarwal, Anil, *Dying Wisdom*, New Delhi: Centre for Science and Environment, 1999.

Becchetti, Leonardo, Nazaria Solferino, and M. Elisabetta Tessitore, 'How to Safeguard World Heritage Sites? A Theoretical Model of "Cultural Responsibility"', University of Rome Centre for Economic and International Studies Tor Vergata Research Paper Series, 12: 6; No. 318, July 2014.

Bertacchini, Enrico, Giangiacomo Bravo, Massimo Marrelli, and Walter Santagala, 'Defining Cultural Commons', in E. Bertacchini, G. Bravo, M. Marrelli and W. Santagata (eds), *Cultural Commons: A New Perspective on the Production and Evolution of Cultures*, Cheltenham: Edward Elgar Publishing, 2012, pp. 3–18.

Bertacchini, Enrico, Claudia Liuzza, Lynn Meskell, and Donatella Saccone, 'The Politicisation of UNESCO World Heritage Decision Making', *Public Choice*, 2016, 167(1): 95–129.

Bertacchini, Enrico, Donatella Saccone, and Walter Santagata, 'Embracing Diversity, Correcting Inequalities: Towards New Global Governance for the UNESCO World Heritage', *International Journal of Cultural Policy*, 2011, 17(3): 278–288.

Brancaccio, Pia, 'World Heritage Sites of Ajanta, Ellora and Elephanta', paper presented at conference on Heritage in Context: Balancing the Global With the Local, New Delhi, 20–22 August 2016 (Chapter V in this volume).

Buzio, Aldo, and Alessio Re, 'Cultural Commons and New Concepts Behind the Recognition and Management of UNESCO World Heritage Sites', in E. Bertacchini, G. Bravo, M. Marrelli and W. Santagata (eds), *Cultural Commons: A New Perspective on the Production and Evolution of Cultures*, Cheltenham: Edward Elgar Publishing, 2012, pp. 178–193.

Centre for Science and Environment, *State of the Environment Report: Traditional Water Systems*, 4th edition, New Delhi: Centre for Science and Environment, 2003.

Center for Science and Environment, 'Traditional Rural Rainwater Harvesting', http://www.rainwaterharvesting.org/Rural/Traditional.htm (accessed on 26 June 2018).

Comptroller and Auditor General of India (CAG), 'Performance Audit of Preservation and Conservation of Monuments and Antiquities of Union Government', Report No. 18, 2013, Ministry of Culture.

Frey, Bruno, and Paolo Pamini, 'Making World Heritage Truly Global: The Culture Certificate Scheme', University of Zurich Institute for Empirical Research in Economics Working Paper Series, ISSN 1424-0459, No. 419, June 2009.

Guiso, Luigi, Paola Sapienze, and Luigi Zingales, 'Does Culture Affect Economic Outcomes', *Journal of Economic Perspectives*, 2006, 20(2): 23–48.

Hayes, Tanya, 'Parks, People, and Forest Protection: An Institutional Assessment of the Effectiveness of Protected Areas', *World Development*, 2006, 34(12): 2064–2075.

Hess, Charlotte, 'Constructing a New Research Agenda for Cultural Commons', in E. Bertacchini, G. Bravo, M. Marrelli and W. Santagata (eds), *Cultural Commons: A New Perspective on the Production and Evolution of Cultures*, Cheltenham: Edward Elgar Publishing, 2012, pp. 19–35.

Labadi, Sophia, and Peter G. Gould, 'Sustainable Development: Heritage, Community, Economics', in Lynn Meskell (ed.), *Global Heritage: A Reader*, 1st edition, New York: John Wiley, 2015, pp. 196–216.

Lal, Pranay, *India: A Deep Natural History of the Indian Subcontinent*, New Delhi: Random House India, 2016.

Lasania, Yunus, 'Badi Baoli Provides a Silver Lining', *The Hindu*, 6 May 2016.

Meskell, Lynn, 'World Heritage and WikiLeaks: Territory, Trade and Temples on the Thai-Cambodian Border', *Current Anthropology*, 2016, 57(1): 72–95.

Meskell, Lynn, C. Liuzza, E. Bertacchini, and D. Saccone, 'Multilateralism and UNESCO World Heritage: Decision-Making, States Parties and Political Processes', *International Journal of Heritage Studies*, 2015, 21(5): 423–440.

O'Brien, D., *Measuring the Value of Culture*, London: Department for Culture, Media and Sport, 2010, pp. 22–33.

Ostrom, Elinor, 'Beyond Markets and States: Polycentric Governance of Complex Economic Systems', Nobel Prize Lecture, Stockholm, 8 December 2009.

Pathak, Subhash, 'Mauryan Irrigation Revives Dry Magadh', *Hindustan Times*, 4 January 2018.

Rajaraman, Indira, 'Climate Change and Other Wishes', *Mint*, 8 January 2016.

Rajaraman, Indira, 'Valuation of World Heritage', India International Centre Occasional Paper Series, No. 17, New Delhi, December 2016.

Ray, Himanshu Prabha, 'From Multi-Religious Sites to Mono-Religious Monuments in South Asia: The Colonial Legacy of Heritage Management', in Patrick Daly and Tim Winter (eds), *Routledge Handbook of Heritage in Asia*, London and New York: Routledge, 2012, pp. 69–84.

'Report of the Thirteenth Finance Commission', December 2009.

Schlager, Edella, and Elinor Ostrom, 'Property Rights Regimes and Natural Resources: A Conceptual Analysis', *Land Economics*, 1992, 68(3): 249–262.

UNESCO, *Operational Guidelines for the Implementation of the World Heritage Convention*, Paris: UNESCO, 2011.

UNESCO, *World Heritage Committee Fortieth Session, Istanbul, Turkey, 10–20 July 2016*, Paris: WHC/16/40.COM/14, 27 May 2016 and WHC/16/40.COM/15, 10 June 2016.

Van Blarcom, Brian Logan, and Cevat Kayahan, 'Assessing the Economic Impact of a UNESCO World Heritage Designation', *Journal of Heritage Tourism*, 2011, 6(2): 143–164.

Appendix 1
Criteria for World Heritage recognition

For recognition as a cultural site, the World Heritage Committee must find that it meets one or more of the first six criteria, and for a natural site, one or more of the last four.

1 to represent a masterpiece of human creative genius;
2 to exhibit an important interchange of human values, over a span of time or within a cultural area of the world, on developments in architecture or technology, monumental arts, town-planning or landscape design;
3 to bear a unique or at least exceptional testimony to a cultural tradition or to a civilisation which is living or which has disappeared;
4 to be an outstanding example of a type of building, architectural or technological ensemble or landscape which illustrates (a) significant stage(s) in human history;
5 to be an outstanding example of a traditional human settlement, land-use, or sea-use which is representative of a culture (or cultures), or human interaction with the environment especially when it has become vulnerable under the impact of irreversible change;
6 to be directly or tangibly associated with events or living traditions, with ideas, or with beliefs, with artistic and literary works of outstanding universal significance. (The Committee considers that this criterion should preferably be used in conjunction with other criteria);

For natural heritage sites, the following criteria are considered:

7 to contain superlative natural phenomena or areas of exceptional natural beauty and aesthetic importance;

8 to be outstanding examples representing major stages of earth's history, including the record of life, significant ongoing geological processes in the development of landforms, or significant geomorphic or physiographic features;

9 to be outstanding examples representing significant ongoing ecological and biological processes in the evolution and development of terrestrial, fresh water, coastal and marine ecosystems and communities of plants and animals;

10 to contain the most important and significant natural habitats for in-situ conservation of biological diversity, including those containing threatened species of outstanding universal value from the point of view of science or conservation.

Source: http://whc.unesco.org/en/146 (last accessed on 9 May 2018)

Appendix 2
World Heritage Sites in India

Cultural	Year of World Heritage status		Natural	Year of World Heritage status
Caves				
1. Ajanta, Maharashtra	1983	1.	Kaziranga Sanctuary, Assam	1985
2. Ellora, Maharashtra	1983	2.	Manas Sanctuary, Assam	1985
3. Elephanta, Maharashtra	1987	3.	Keoladeo National Park, Rajasthan	1985
4. Bhimbetka Rock Shelters, Madhya Pradesh	2003	4.	Valley of Flowers, Uttarakhand	1988
Large sites, multiple structures		5.	Sunderbans Estuarine Mangroves, West Bengal	1987
5. Mahabalipuram Rock Temples, Tamil Nadu	1984	6.	Western Ghats, 39 sites in four states	2012
6. Hampi ruins of Vijayanagar kingdom, Karnataka	1986	7.	Great Himalayan National Park, Himachal Pradesh	2014
7. Fatehpur Sikri, Uttar Pradesh	1986			
8. Khajuraho temples, Madhya Pradesh	1986			
9. Pattadakal Monuments, Karnataka	1987			
10. Sanchi Buddhist Monuments, Uttar Pradesh	1989			

Cultural	Year of World Heritage status	Natural	Year of World Heritage status
11. Champaner-Pavagadh Archaeological Site, Gujarat	2004		
12. Nalanda Mahavihara, Bihar	2016		
Single Monuments		Cultural currently functioning	
13. Humayun's Tomb, Delhi	1993	22. Churches and Convents of Goa	1986
14. QutabMinar, Delhi	1993	23. Mahabodhi Temple Complex, Bodhgaya, Bihar	2002
15. Agra Fort, Uttar Pradesh	1983	24. Shivaji Railway Terminus, Maharashtra	2004
16. Jantar Mantar, Rajasthan	2010	25. Chola Temples, Tamil Nadu	1987
17. Taj Mahal, Uttar Pradesh	1983	26. Mountain Railways, three sites	1993-2008
18. Hill Forts, Rajasthan	2013	27. Le Corbusier structures at Chandigarh	2016
19. Rani-ki-Vav, Gujarat	2014	28. Historic City of Ahmedabad	
20. Konarak Sun Temple, Orissa	1984	Mixed Cultural and Natural	
21. Red Fort, Delhi	2007	29. Khangchendzonga National Park, Sikkim	2016

Source: http://whc.unesco.org/en/list/ (last accessed on 9 May 2018)

Appendix 3

Funding of forest grant from national to state governments in India, 2010–2015

Green grants: Forest Protection [Rs. 5,000 crore (2010–2015)]

A reward in the structure of statutory federal transfers from the national government to state governments for the ecological and bio-diversity externalities generated by standing forests falling within their jurisdictions. It was designed as compensation to states who perceive very few benefits internal to their jurisdictions relative to the economic disability of area under forests. The formula follows.

Where G_i is the share for state i in the forest grant:

$$G_i = [Num_i] / \sum_{i=1}^{n} [Num_i], \text{ where } Num_i$$

$$= [\{\frac{F_i}{\Sigma F_i} + R_i\} * \{1 + ((M_i + 2H_i) / A_i)\}]$$

Where
F_i, M_i, H_i are total, moderately dense, and highly dense, forest area of state i
A_i is the total area of state i

$$R_i = max \, [0, \{\frac{F_i}{A_i} - \frac{\Sigma F_i}{\Sigma A_i}\} / 100]$$

The total grant accruing to each state was to flow as follows:

- 25% over 2010–2012, to cover costs of making working plans for all forest divisions within the state.

• 75% over 2012–2015, pro-rated to the number of forest divisions with approved working plans. In terms of usage of the amount released, 25% is earmarked for forests, with the rest usable for any developmental purpose.

The source for area under forests used to define state shares were all from the State of the Forest Report 2009, whose data pertained to the year 2007.

In addition to the forest protection grant, the Finance Commission recommended state-specific grants, not determined by formula, but prescribed for states as absolutes and not as a percentage of any targeted sum. The state-specific purposes varied widely, but included preservation of particular water bodies and wildlife reserves and upkeep of heritage sites and archaeological properties not under the purview of the Archaeological Society of India. Around 60% of the total set aside for these purposes was for cultural heritage, and 40% for natural heritage.

Source: Report of the Thirteenth Finance Commission, December 2009.

4 The multivalence of landscapes

Archaeology and heritage

Uthara Suvrathan

Of all the social sciences, Archaeology is one that has taken strong root in the public imagination. However, there is very often little dialogue between the academic understanding of an archaeological site and the local, lived experience of it. This chapter is an attempt to grapple with these different levels of engaging with the landscape without privileging either.

At one level, local communities are seen as repositories of transmitted historical wisdom, unchanging and essential. This belief has structured considerable archaeological investigation in south Asia ranging from the village-to-village surveys of the Archaeological Survey of India (ASI) to the hiring of "local" guides and informants by most current archaeological projects. Once a site is identified, the archaeologist takes over, changing the local narrative to an academic one with little more than a tip of the hat to traditional histories, which are barely acknowledged in brief contextual introductions to detailed archaeological reports filled with a plethora of beautifully organised and illustrated data on site sizes, pottery styles, architectural forms and so on. It is no wonder that such reports, so fascinating to archaeologists, find little favour with the public at large.

How do we bridge this gap? I often wonder whether we as archaeologists end up essentialising the past. In our attempts to reconstruct past lifeways we frequently tend to create sanitised images that ignore the messiness and disorganisation of real life. However, a critical examination of the archaeological landscape provides interesting insights into the ways people in the past (and in the present) engaged with the built and natural environment. In addition, a critical understanding of modern cultural practices as centred on ancient archaeological sites can provide interesting insights into patterns of change – particularly those of remembrance, forgetting and erasure.

In this chapter, I draw upon archaeological work that I have been involved in over the last several years, highlighting the multiple layers that go into the making of historical sites. I move from the academic – in this case archaeological – construction of historic sites to examining the lived experience of the community, past and present. I conclude by briefly commenting on the modern impetus towards cultural heritage management and the uneasy relationship with academic approaches.

Archaeology in Banavasi: rethinking dynasties

Over the past several years, I have been conducting archaeological research in Banavasi, an ancient urban centre located in the modern state of Karnataka. Banavasi is a particularly fascinating site to study since it was the locus of regional elite individuals or families who maintained complex relationships of allegiance or opposition to numerous cycling states and empires that controlled large parts of southern India from the late centuries BCE to the 18th century CE. While Banavasi remained on the peripheries of these larger political systems for much of its occupation, between the 4th and 7th centuries CE it was itself at the centre of the regional kingdom of the Kadamba lineage.

Previous archaeological research at Banavasi in the 1960s and 1970s was conducted with the aim of understanding the earliest periods of occupation at the site, specifically the "Early Historic", broadly dated to between the late centuries BCE and the 6th century CE in southern India. In addition, there was a focus on tracing the archaeological correlates of specific ruling dynasties, in particular the Satavahanas and the Mauryan emperor, Ashoka, and on identifying the role of Buddhism at the site.[1] This archaeological emphasis on particular moments in history, whether dynastic or religious, reflects the culture-historical foundations of archaeology in the subcontinent, and is a holdover from the earliest phase of colonial archaeology when Alexander Cunningham wandered the land, following in the footsteps of Chinese pilgrims to the subcontinent in order to identify ancient Buddhist monastic and religious sites.

The implicit quest for the presence or absence of specific (royal) elite groups, who were considered central to the historical development of a region, especially if they were mentioned in textual sources, is problematic to say the least. It is not surprising therefore that peripheral places remain poorly understood, in part due to this archaeological preoccupation with "kings and their things" and the resultant focus on the establishment and disintegration of large states and empires in what have been considered "core areas", as well as the assumption that these complex systems played a determinant and causal role in developments outside the cores, such as in peripheral places like Banavasi. Through my work I question some of these assumptions by showing that the archaeological study of the landscape provides a deep temporal perspective that allows us to reconstruct long-term historical patterns of continuity and change in peripheral regions outside of or on the edges of larger political systems.

A central aim of my research at Banavasi was therefore to interrogate the sources of our implicit and often unconscious "dynastic thinking" and move from a "Big man" version of historical process where Banavasi's history is one punctuated by the rise and fall of ruling dynasties to one that takes into account long-term processes of continuity and change, both of external influences and local developments. Between 2009 and 2012, I directed the Banavasi-Gudnapura Regional Survey Project (BGRS) that investigated the organisation of regional socio-political complexity at Banavasi and at the adjacent contemporaneous centre of Gudnapura through the systematic archaeological survey of a 50 sq km area block centred on the two sites. During this project, more than 300 historical sites were recorded and analysed, each given a unique numeric identifier.[2]

From an archaeological standpoint, the materiality of the landscape and the physical manipulation of space through construction, inscription and even destruction helped create and define the environment in which people lived, and revealed considerable information about beliefs and practices at particular points in history. I focused broadly on a time period from the late centuries BCE to around the 18th century to examine the complex of historical structures and spaces that constituted the archaeological landscape in Banavasi and its hinterland. Through a Geographic Information Systems (GIS)–based analysis of the landscape, I was able to reconstruct complex patterns of historical development to identify three broad temporal phases in Banavasi's history, each characterised by unique spatial configurations in the archaeological landscape and therefore of socio-economic, political and religious organisation. The three periods spanned the late centuries BCE to the 7th century CE, the 7th century CE to the 14th century and finally, the 14th century to the 18th century (Figure 4.1).

Although located in a peripheral area, Banavasi can be considered a regional centre of some permanence – a "peripheral core". Banavasi's development and longevity, despite the area's incorporation within successive larger imperial political systems, can be linked to two main factors: the presence of local elite families that were closely connected

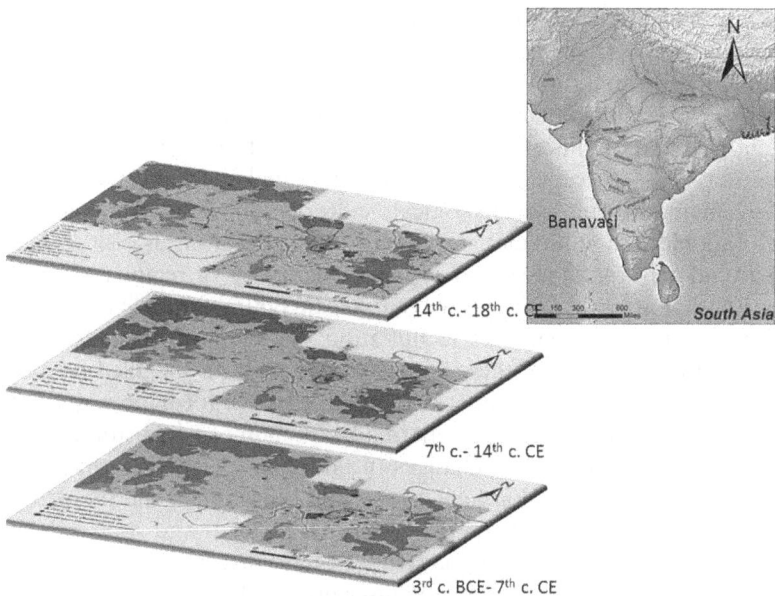

Figure 4.1 Landscape change: Banavasi, c. 3rd century BCE–18th century CE.

Source: Courtesy of the author

to the area, and to the development of a complex relationship of patronage and legitimacy between elite and religious groups (especially mainstream Brahmanical Shaivism) that played out in the environs of Banavasi through the manipulation of the landscape.

Banavasi's initial development during the late centuries BCE and early centuries CE can be placed within the context of larger subcontinent-wide developments when it was one of several nodal points along a long-distance trade and Buddhist pilgrimage network. That some form of elite authority was concentrated at Banavasi from a very early period is apparent from the fact that the early settlement was fortified. In South India during this period, several elite families, including the Chutus, likely bolstered their authority by establishing a pattern of donation to Buddhist institutions. At Banavasi too, this period sees the establishment of numerous Buddhist stupas in close proximity to the city, a pattern that replicates that seen elsewhere in the subcontinent.

Subsequently, Banavasi's rise to prominence as a regional capital can be linked to the presence of a substratum of regional elite families that strategically appropriated the Banavasi area as one of their core areas. While they are not visible archaeologically at Banavasi, between the 4th and 7th centuries CE, numerous inscriptions issued by the Kadamba family are found across the modern state of Karnataka and testify to the establishment of a regional kingdom. Through my research I showed that, significant though they may have been, dynastic changes cannot necessarily be correlated to artifactual markers. Conflating dynastic names and archaeological "cultures" subsumes the diversity of regional and sub-regional patterns of organisation as reflected in material culture, and leads to explanatory models of regional organisation that assign a primary role to dynastic politics and the establishment of states and empires in initiating socio-political change both within the core areas of these polities and outside, in so-called peripheral places. At the same time, I recognise that the formation, growth and decline of dynasties represent important moments in history, often indicative of fundamental changes in political, social, economic and religious organisation. I argue however, that the impact and trajectory of these developments is unique to specific regional and sub-regional contexts, as in the case of Banavasi. As I have discussed elsewhere,[3] a more critical consideration of the inscriptions of the early Kadambas highlights the complex and fragmented nature of their authority. In fact, a reinterpretation of these early inscriptions does not reveal a near unbroken line from father to son but dynamism and constant reorganisation of royal elite power. Ultimately, the inscriptions also show that the Kadamba family successfully established its kingdom through a political strategy that involved a complex pattern of gift-giving to Brahmanical religious groups linked to legitimacy.

Post 7th century, Banavasi's continued importance lay partly in its establishment as a regional sacred and pilgrimage centre for the Brahmanical Shaivite tradition. This involved a relationship of elite patronage and Brahmanical legitimation enacted through elite patronage of temples and of the grants of land to Brahmanical individuals and institutions. The early Buddhist landscape at Banavasi was now completely replaced by a Hindu one, with numerous temples being constructed within and around the city. Banavasi itself, and especially the central temple, thus provided a space for the creation, maintenance and display of elite authority and alliances at various levels in the hierarchy through these grants that often mapped a hierarchical geography of power, placing each donee with respect both to local families and to the imperial rulers under whose aegis they ruled. It is important to note here that the majority of inscriptions and grants of land are not by the imperial rulers who controlled Banavasi and its vicinity but by regional elite groups owing allegiance to them. In traditional historiography, regional organisation in this late period has been considered to be "feudal", with a proliferation of intermediate groups that, it is argued, aggravated and were symptoms of the fragmentation of centralised authority. However, while the large numbers of inscriptions recording grants of land by local groups is certainly noteworthy, in the Banavasi area at least, the presence of these regional elite groups was not a sudden development. In fact, these groups, such as the Kadambas, might best be considered the stable entities in this region.

The archaeological landscape at Banavasi reflects the complex pattern of interaction with larger political systems, as well as the continuity of local elite groups. There is, for instance, evidence for the increasing penetration of a number of empires in the region post 10th century, as reflected in temple architectural styles. At the same time, there was continuity in local traditions and architectural styles, as well as the local organisation of agricultural resources and the small-scale, localised production of laterite, brick and tiles. This local organisation can be linked to the continued importance of local elite families in the region. The presence of these elite groups probably ensured a certain level of continuity and stability while the region cycled in and out of the control of successive imperial powers. It is only in the final phase post 14th century that there are some increasing signs of competition and conflict, both religious and political as there is evidence for the increasing influence of an imperial architectural and religious idiom, the rise of new anti-Brahmanical religious movements and competition from neighbouring settlements.

A central focus of my work therefore was to address macro-level questions of regional organisation – to think about polity and patronage, authority and power, and a complex and ever-changing hierarchy of political and religious power over the long term. In my work I questioned the idea that peripheral regions were necessarily static entities. Instead, against a historical context of cycling and ephemeral states and empires, smaller but long-lived peripheral areas characterised by small centres shaped by the presence of elite families (and not necessarily dynastic lineages) are essential units of historical analysis.

Thus far, I have talked about changing academic perspectives on Banavasi – from an interest in its position as a Buddhist site and as the Kadamba capital to a consideration of longer term processes and in the role of local elite and non-elite groups. On the one hand, the academic perspective is one of the macro-scale, attempting to address larger issues of culture change, religious influence, political organisation and so on. On the other hand, we have the more personal experience of material culture – the nuts and bolts of archaeology – as a non-academic and local audience interacts with and experiences the material culture that we so avidly study as archaeologists. We tend to ignore this more personal and immediate aspect of archaeology, and I have been guilty of this too; the following comments are more in the nature of thinking aloud about the micro-scale.

In the following sections I want to draw attention to the lived experience of the landscape, both past and present – a perspective that adds considerable richness to the macro narratives so summarily reviewed earlier. There has been considerable academic interest in the more prominent archaeological sites and structures at Banavasi. However, these sacred and secular structures themselves helped create and define the environment in which the inhabitants of the area lived – and yet, at the same time they were and are constantly reinvented, playing different roles throughout time. In particular, I highlight several interlinked themes in this constant reinvention of the landscape at Banavasi: perception; memory (both in terms of remembrance and forgetting); replacement and reuse; life-history/reinterpretation and destruction.

Perception: Buddhist Banavasi, past and present

Buddhism at Banavasi exemplifies something of the complex nature of landscape as created, both physically through the manipulation of the landscape as well as ideationally, through the creation of imagined landscapes and theoretical constructs that attempt to define the nature and characteristics of a particular landscape.

Much of the academic interest in Banavasi stems from our perception of the site as an important Buddhist site and has been greatly shaped by information derived from textual traditions that construct an imagined landscape of Buddhist pilgrimage and patronage, a connection that is also borne out by some early inscriptions. The Sri Lankan Buddhist chronicles in Pali dating to the early centuries CE (the *Dipavamsa* and *Mahavamsa*) record that, following a Buddhist council at Pataliputra, missionaries were dispatched to several regions, including Vanavasi (an alternative term for Banavasi used in several early texts).[4] Three early inscriptions also inform on the presence of Buddhist practice at Banavasi. First, a Prakrit Ikshvaku inscription from Nagarjunakonda in the modern state of Andhra Pradesh dated to the 3rd or 4th centuries CE records the visit of Buddhist monks from Sri Lanka to Vanavasi among many other regions.[5] Second, another inscription from the same place and period records the construction of a vihara by Kodabalasiri, daughter of Sri Virapurushadatta, sister of the (Ikshvaku) King Vashishthiputra Sri Ehuvula-Chamtamula, and the queen of an unnamed king of Vanavasaka.[6] The third inscription is from Karle in modern Maharashtra, and records the construction of a *shilagraha* (stone chamber) by a merchant from Vaijayanti.[7]

While the memory of a Buddhist Banavasi has been transmitted to us via texts and inscriptions, archaeological evidence for Buddhism is limited. A 2nd century CE inscription found at Banavasi commemorates the queen of King Vashishthiputra Shiva Sri Satakarni. This inscription is recorded on a rectangular slab with an eroded *chaitya* (a Buddhist shrine, typically with a stupa at one end) motif on the top.[8] A 3rd century CE inscription records the construction of a tank, *vihara* (Buddhist monastery) and *naga* (a deity in the forma of a snake, often associated with prosperity and fertility) by Nagashri, the daughter of the Chutu king, Vishnuskanda Satakarni, at Banavasi. The early excavations at Banavasi also uncovered two apsidal brick structures that are very similar to Buddhist chaityas found elsewhere in the subcontinent.[9] However, there is no conclusive evidence (images or inscriptions) that these structures were associated with Buddhist worship.[10] On survey, we identified ten possible stupa sites due to their distinctive circular shape and evidence that they were structural mounds (Figure 4.2). All have some evidence of brick construction and a few have scatters of flat red tiles with distinctive shapes that date them to

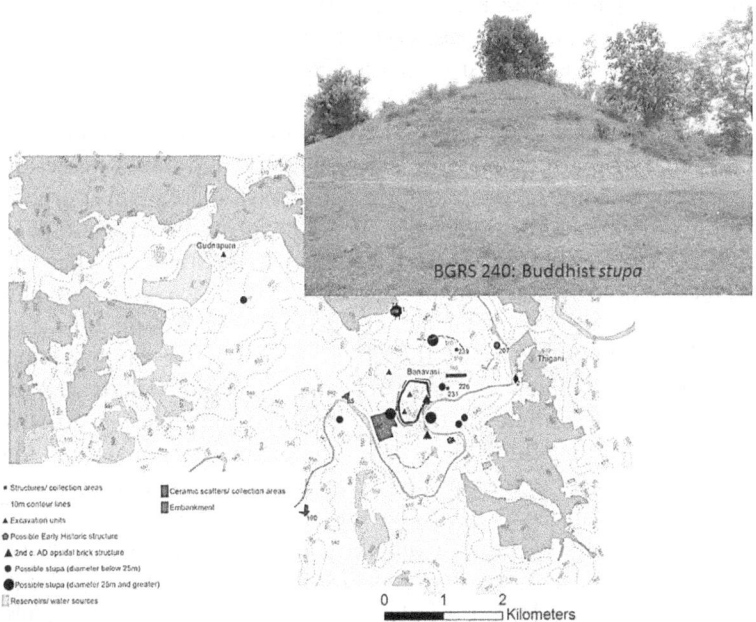

Figure 4.2 Buddhist stupas at Banavasi.

Source: Courtesy of the author

the early centuries CE. Since these sites have not been excavated and most are overgrown or disturbed by agricultural activities, their identification is necessarily tentative. This limited evidence for Buddhism at the site does not quite fit the textual imaginary of Banavasi as an important Buddhist centre.

A close connection between Buddhism and political authority has been discussed in several studies.[11] In some parts of the subcontinent, in fact, Buddhism's introduction in the 3rd century BCE has been considered to be the catalyst that supported the transformation of 'undifferentiated societies into a single powerful kingly polity'.[12] In the Banavasi area, however, while some of the pre-Kadamba elite families might have patronised Buddhism, there is little evidence for early Kadamba patronage of this religion. The vast majority of their inscriptions issued between the 4th and 7th centuries CE reference support to Brahmanical institutions. This likely contested nature of interaction between Buddhism and political authority is also borne out by the fact that most of the stupa sites are located near Banavasi but outside of the main settlement, a pattern that is frequently noticed elsewhere in south Asia.

In today's Banavasi, Buddhism no longer holds an important place in current memory and the erstwhile Buddhist landscape is rapidly being appropriated. Rarely are the remnants of the stupa mounds recognised as Buddhist structures, but are instead more or less forgotten or are absorbed into a modern mythos that weaves tales of ancient mounds or *guddas* that were the palaces of ancient (and unnamed) kings. One of the stupas has in fact been converted into a makeshift Hindu shrine comprised of sculptural fragments likely appropriated from a temple placed on the flattened top of a stupa mound (Figure 4.3). Yet another, BGRS 207, was a possible stupa site that was completely destroyed to expand agriculture: there we were only able to record the outlines of the site and collect a few sample tiles (Figure 4.3). When we re-visited the site a year later, the site had more or less been completely erased as repeated ploughing obliterated the original circular outline of the structure and pulverised distinctive ancient tiles to unidentifiable pieces. One of the few remaining fragments of memory of a Buddhist Banavasi lies in the fact that several of the mounds identified as stupas have looter's holes on the top. From colonial travellers accounts from the 17th and 18th centuries, we know that the topes were often mined for reliquaries by the straightforward method of digging a hole in the top down to the relic chamber. While the looter's holes in the Banavasi stupas cannot be dated, it is an interesting remnant of a memory or belief that there might be *"treasure"* in the centre of these structures.

Figure 4.3 Buddhism forgotten: appropriation and destruction.

Source: Courtesy of the author

Memory

In addition to the rapidly fading memory of a Buddhist Banavasi, the area around the city is alive with meaning and a complex net of stories, myths and connections. One of the most common forms of memorialisation inscribed on the landscape are a range of memorial stones that commemorate the life and death of a variety of individuals, and date roughly between the 8th to the 16th centuries or later. Hero stones (*viragals*) were carved rectangular slabs that recorded the death of a "hero" in war on behalf of a king, or in defence of cows, women or the village, or in self-sacrifice for various reasons. The most common type recorded during our survey has three panels: the lowest contains elaborate depictions of the hero in battle, the middle shows the hero flanked on either side by *apsaras* (celestial women), and the topmost panel depicts him in heaven or worshipping a Shivalinga (representing his location in Kailasha or heaven). In several cases, the spaces between the panels contain inscriptions describing the death of the hero and recording any grant of land made in favour of the family of the deceased. Simpler forms included single-paneled versions of this more elaborate form. This form of commemoration drew heavily on established imperial artistic styles and clearly displayed an elite heroic ideal. The proliferation of these forms of commemorating death in the post-8th century CE period also coincides with increasing political flux as the Banavasi area was periodically incorporated into larger political systems. I suggest that these stones may be indicative of a period of increasing warfare and conflict.

Sati stones (*mahasatikal* or *mastikal*) commemorated the immolation of a woman on the death of her husband. Perhaps the most common is a stylised representation consisting of a long pillar with a woman's adorned arm extending from it. A full pot (*purna-kumbha*) is shown on top of the pillar, depicting abundance and a sign of auspiciousness. Less common were crudely carved male and female figures against a plain background, the woman often with her right arm raised as in the first type of *sati* stone. These latter figures are at times crudely carved and stylistically seem to draw from a local tradition and not from the more polished pan-Karnataka sculptural tradition. The female figures wear elaborate headdresses and wide, pleated skirts and not the elaborately draped garments so often depicted in Chalukya and Rashtrakuta art and can be dated to a post-16th century period (Figure 4.4).

Figure 4.4 Hero and Sati stones.

Source: Courtesy of the author

Jaina memorial stones (*nishidhi*) commemorated the death of a Jaina ascetic through the rite of *sallekhana,* involving detachment from the world ending with a fast to death, and was a common practice between the 6th/7th centuries CE and the 15th century.[13] The simplest form of memorialisation to such individuals was a slab depicting the footprints of the deceased in relief, and similar forms have been dated as early as the 6th and 7th centuries CE.[14] From the 8th/9th centuries onwards, *nishidhi* stones took on some of the stylistic features of hero stones, with two or three panels showing the life and after-life of the ascetic. These panels showed the ascetic in heaven or leaving his body, and scenes from the life of the ascetic, including panels where he is shown as listening (often with his wife) to the teachings of his guru (Figure 4.5).

A final form of commemoration that marks the landscape are simple, often crudely carved stones with inscribed or bas-relief depictions of a linga frequently with a seated cow next to it (Figure 4.5). Often there is a carved sun and moon on the top left and right respectively, indicating that the gift is eternal, a standard motif in post-12th-century carved stones in Karnataka.[15] These stones are known locally as *linga-mudra*

Linga-mudra Stone

Nishidi Stone

Figure 4.5 Nishidi and *linga-mudra* stones.

Source: Courtesy of the author

stones; while none of them depict a human worshipper, it is likely that they are some form of commemoration, whether of death or of devotion is unclear. Stylistically, these stones show great variation, are rarely inscribed, and seem to represent a limited investment of resources since they are generally poorly finished and crudely carved. I suggest that they might represent non-elite forms of commemoration or devotion located as they are scattered away from settlements in agricultural fields. In other contexts, these stones have been identified as boundary markers in agricultural fields (Sinopoli, personal communication).

These memorials clearly mapped a pre-modern landscape of religious and secular memory. In the post-8th century CE period, the Shaiva tradition became increasingly important (this is borne out by the construction of temples in the area as well) and most of the hero stones have a clear Shaiva affiliation. The Jaina tradition also had a presence in the region but the memorials are much less numerous. Today, long after the memories of the individuals in whose honour they were erected have faded, these stones continue to be a part of the present-day religious landscape. *Sati* stones for instance are interesting in that they have a sanctity of their own, without any clear association with mainstream Brahmanical temples. They are typically located outside temple compounds or in their own smaller shrines, unlike the hero stones that are usually placed right next to or within the temple compound. Also unlike the hero stones, *sati* stones are worshipped even today as "Masti-amma", a local form of the goddess.

Replacement and reuse

Finally, I want to talk about the constantly changing nature of the landscape around Banavasi as exemplified by processes of replacement and reuse. Due to the academic tendency to classify temples by the initial dynastic period of their constructions ("Kadamba period", "Vijayanagara temple"), we tend to forget that they were complex entities that had life histories long after that of their initial construction as they were constantly added on to and altered. For instance, the central Madhukeshwara temple in Banavasi was constantly added on to after its original construction in the 7th or 8th centuries CE, with major elaborations in the 11th, 14th and 16th centuries. In fact, during the past seven years that I was in and out of Banavasi there have been constant small architectural and artistic additions to the central structure. Not only are religious structures added on to during their long life spans but there are also instances where structures were appropriated by other traditions. For instance, a 12th-century Jaina *basadi* in the neighbouring village of Gudnapura was converted to a

Hindu Virabhadra temple, likely in the 16th century. Subsequently, the structure fell out of worship and was abandoned. Recently, the ASI has been conducting extensive excavations in the area immediately around the structure, purportedly uncovering a palace–temple complex dating to the Kadamba period or the 5th to 6th centuries CE.

Yet another example of these processes of reinterpretation and reuse comes from a consideration of folk and local religious practices that often occur outside of the traditional ritual spaces of the temple and have a complicated relationship with the dominant Brahmanical tradition (Figure 4.6). Instead of looking at the multiplicity of religious traditions solely in terms of conflict or even necessarily of incorporation (although both occur), we can see Banavasi as a stage where the availability of resources and patronage allowed for shared religious space. The present landscape in the survey area, dotted as it is with small, aniconic shrines, *naga* (snake) stones and other sacred spaces that do not always have a constructed component is a constant reminder of the continuity and presence of local religious practice.

Throughout South India, folk beliefs populate the landscape with a variety of divine and semi-divine beings, as well as spirits (*bhutas*) and other inimical forces. Landscapes thus comprise not only

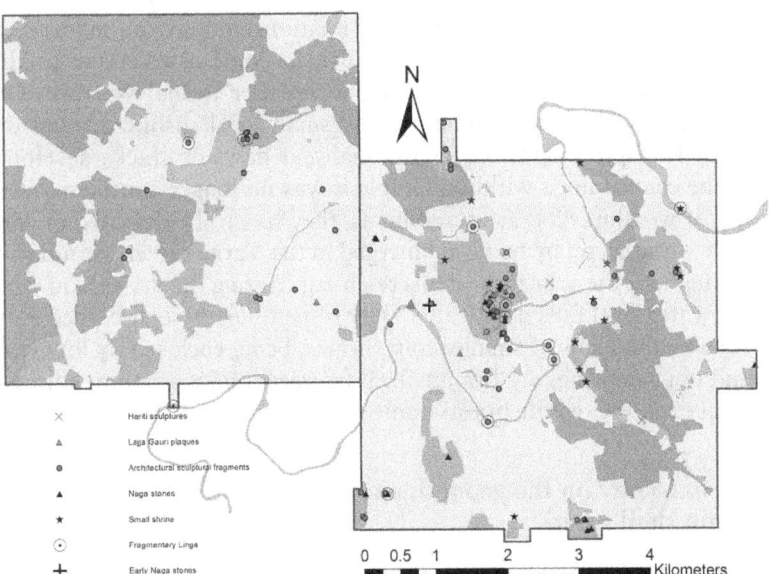

Figure 4.6 Archaeological landscape in the survey area around Banavasi.

Source: Courtesy of the author

constructed spaces but "invisible" ones as well.[16] In many cases, the highly visible constructed landscape of Brahmanical temples is interspersed with numerous smaller sites that are often "visible" to the inhabitants of that landscape but less visible or even invisible to an outsider. Folk shrines or sacred places in the landscape that are not obviously marked but are known and have importance include a range of rounded stones or earthen pots worshipped as one of the forms, the goddess *Chowdamma*; places identified as residences of spirits and various natural features (termite mounds, snake holes, banyan trees).

In other cases, these shrines can include miscellaneous architectural or sculptural fragments appropriated from larger structures. These ephemeral forms of construction are a crucial part of the wider religious landscape and are as important in lived practice as the larger stone temples. Such small village shrines are simply made of easily available materials and require little labour. Due to their very impermanence, they are continuously maintained – cleaned, added to, worshipped. These small shrines are a more organic feature of the village landscape (a temple lintel tucked away under a banyan tree, a broken sculpture under a palm leaf shed). I cannot imagine that such places would leave easily identifiable traces for the archaeologist, and yet they must have been a part of village life from time immemorial.

The life histories of some of these shrines are extremely interesting, showing as they do a complex relationship between mainstream Brahmanical tradition and local cults. In some cases, these smaller shrines travel complex trajectories to re-enter formal worship. One such example is the proliferation of small local shrines around discarded temple fragments. Traditionally, if flaws or cracks developed in the central linga within a temple it was no longer considered worthy of worship. Yet, as sacred items, they had to be disposed of carefully – and were, by being submerged in the Varada river. Periodically throughout the year these items re-emerge during the dry season when the water level falls drastically. Over some time, these discarded items become the focus of smaller folk shrines, being enclosed by low walls and not requiring any Brahmanical intercession in worship, although frequently the temple priests come to make offerings at the site.

Archaeology on the ground: and ne'er the twain shall meet?

The historical and modern landscape at Banavasi is comprised of numerous layers of meaning and constant interpretation and reinterpretation. I have talked so far about the diverse ways in which different groups understand and experience the landscape – from changing academic approaches to a more intimate lived experience of the landscape. This discussion will however be incomplete without some consideration of present-day local efforts towards cultural heritage management and the complexities of thinking about such a multivalent landscape.

In Banavasi today there is a strong push towards the development of endogenous and sustainable rural tourism.[17] Under the Endogenous Tourism Project–Rural Tourism Scheme (ETP-RTS), a joint scheme undertaken by the Ministry of Tourism of the Government of India and the United Nations Development Programme (UNDP), several initiatives have been undertaken in Banavasi with the involvement of the local Village Tourism Development Committee, Bharathiya Agro-Industries Foundation (BAIF) and BAIF Institute for Rural Development (BIRD). Among others, major infrastructural projects were completed such as the construction of a rest house and of restrooms near the temple.[18] The focus was on capitalising on what are considered to be the key attractions of the area: rural traditions (art, crafts, cuisine), and the natural and cultural heritage. The latter really seems to focus on the importance of the main temple (Madhukeshwara) as a pilgrimage site.

However, while an attempt is being made to write Banavasi into the wider cultural, historical and ecological landscapes of Karnataka, the local archaeological landscape is largely ignored, with the exception of the Madhukeshwara temple. Unfortunately, as is the case in many parts of India, archaeology in what I would call its "raw" form – in its academic avatar – tends to be relatively inaccessible to the public at large. As an example, during our survey the local tourism committee commissioned a Bangalore-based PR group to create a brochure promoting local tourism. One of the members of the team valiantly spent a day surveying with us. Although we talked about what we were trying to do, the day she spent with us was not our most memorable. We walked over 10 kilometres, identified two small ceramic scatters and recorded a small temple. The final brochure, not surprisingly, does not talk about the rich archaeological history and work (including, most importantly, considerable excavation done by the ASI) at Banavasi. The local archaeology and material culture around Banavasi – so important to the lived experience of the area, both past and present – is largely ignored. Instead, the more accessible narrative that is provided is one of the sequence of kings who ruled Banavasi and its place in the larger historical narrative of the region as an important Kadamba site.

There is therefore no accessible version of current academic archaeological research as it pertains to the immediate vicinity of Banavasi for
the wider public. I suggest that one way to bridge this gap is through
the use of artefacts since the very materiality of the past can engage
otherwise lukewarm interest. There had been some interest in the construction of a museum at Banavasi showcasing some of the artefacts
(sculptures and inscriptions primarily) excavated in the area. While a
very nice museum structure has been constructed by the ASI next to
the area of their early excavations, a lack of funds and political issues
between the local, state and central governments have led to the building being abandoned. The numerous artefacts supposed to be housed
in the museum are located in a makeshift storage room in the Banavasi
temple and ignored by tourists in the absence of any descriptive signage. It is of exceeding importance to incorporate the local historical
and archaeological landscape around Banavasi in tourism initiatives.
From an academic perspective it is therefore increasingly relevant for
archaeologists to incorporate public engagement in their projects at
much more organic level than has been done in the area. For instance,
we can learn a lot from the "crowdsourcing" initiatives of the biological sciences where scholars are creatively incorporating public participation in scientific research.

This chapter has traced something of the multiple ways of "seeing"
a site – from changing academic approaches to the local, lived experience of it, to the current heritage management perspective. However,
these varying perspectives are not in productive dialogue since there
are multiple stakeholders and multiple agendas. This exemplifies a
central problem of heritage management projects in complex historical sites such as Banavasi.

Notes

1 A. V. Narasimha Murthy et al. (eds.), *Excavations at Banavasi*, Mysore:
   Directorate of Archaeology and Museums, 1997.
2 Uthara Suvrathan, 'Complexity on the Periphery: A Study of Regional
   Organization at Banavasi, C. 1st–18th Century A.D.', unpublished PhD
   dissertation, University of Michigan, 2013.
3 Uthara Suvrathan, 'Regional Centres and Local Elite: Studying Peripheral
   Cores in Peninsular India', *Indian History*, 2014, 1: 89–142.
4 A. M. Shastri, 'Buddhism in the Deccan During the Sātavāhana Age', in
   Arundhati Banerji (ed.), *Hari Smriti: Studies on Art, Archaeology and
   Indology (Papers Presented in Memory of Dr. Haribishnu Sarkar)*, New
   Delhi: Kaveri Books, 2006, pp. 71–83.
5 J. P. Vogel, 'No. 1 Prakrit Inscriptions From a Buddhist Site at Nagarjunakonda', *Epigraphia Indica*, 1929–30, XX: 7–8, 23.

6 Vogel *op. cit.*, pp. 5–6, 15, 24–25.
7 S. Ritti, 'Buddhism in Kannada Inscriptions', in Devendra Handa and Ashvini Agrawal (eds.), *Ratna-Chandrika: Panorama of Oriental Studies (Shri R. C. Agrawala Festchrift)*, New Delhi: Harman Publishing House, 1989, p. 316.
8 Murthy et al. (eds.), *Excavations at Banavasi*, p. 27.
9 Murthy et al. (eds.), *Excavations at Banavasi*, pp. 67–68.
10 The presence of a Skanda image near one of the structures might indicate a Hindu affiliation. H. P. Ray has suggested that apsidal shrines were part of a religious architectural tradition that dated back to the second century BCE and included Buddhist and Hindu structures as well as those of local cults. Even if these apsidal structures were not Buddhist, they belong to an early, possible pre-fourth century architectural tradition in the region (Himanshu Prabha Ray, 'The Shrine in Early Hinduism: The Changing Sacred Landscape', *The Journal of Hindu Studies*, 2009, 2: 76–96).
11 R. A. E. Coningham, 'Monks, Caves and Kings: A Reassessment of the Nature of Early Buddhism in Sri Lanka', *World Archaeology*, 1995, 27: 222–242; Mark Harrell, 'Sokkuram: Buddhist Monument and Political Statement in Korea', *World Archaeology*, 1995, 27: 318–335; J. Heitzman, 'Early Buddhism, Trade and Empire', in K. A. R. Kennedy and G. L. Possehl (eds), *Studies in the Archaeology and Palaeoanthropology of South Asia*, New Delhi: Oxford University Press, 1984, pp. 121–137; Himanshu Prabha Ray, *Monastery and Guild: Commerce Under the Satavahanas*, New Delhi: Oxford University Press, 1986.
12 Coningham, 'Monks, Caves and Kings', p. 223.
13 S. Settar and Ravi Korisettar, 'Niśidis in Karnataka—a Survey', in S. Settar and G. D. Sontheimer (eds), *Memorial Stones: A Study of Their Origin, Significance and Variety*, Dharwad and Heidelberg: Institute of Indian Art History, Karnatak University and The South Asia Institute, University of Heidelberg, 1982, pp. 283–287.
14 R. Sesha Sastry, 'The Memorial Stones of Karnataka', in D. V. Devaraj and C. B. Patil (eds), *Art and Architecture in Karnataka (Papers Presented at the National Seminar on Archaeology 1985)*, Mysore: Directorate of Archaeology and Museums, 1996, p. 125.
15 Crispin Branfoot and Anna L. Dallapiccola, 'Temple Architecture in Bhatkal and the "Rāmāyaṇa" Tradition in Sixteenth-Century Coastal Karnataka', *Artibus Asiae*, 2005, 65: 259.
16 David Fontijn, 'The Significance of "Invisible" Places', *World Archaeology*, 2007, 39: 70–83.
17 Aditi Chanchani et al. (eds), *Sustainability in Tourism: A Rural Tourism Model*, Bangalore: Ministry of Tourism, Govt. of India/UNDP, 2008.
18 The infrastructural projects have had mixed results. The resthouse is unfortunately not integrated within the village but is located outside the city walls in a fairly isolated and forested area. There is no integrated dining facility and electricity and water are available only for a limited window each day. Does this reflect the complex relationship between local inhabitants and this impetus towards rural tourism? The maintenance of these structures is also extremely spotty. For instance, on a recent visit in 2015, the spanking new restrooms were in terrible shape.

Bibliography

Branfoot, Crispin, and Anna L. Dallapiccola, 'Temple Architecture in Bhatkal and the "Rāmāyaṇa" Tradition in Sixteenth-Century Coastal Karnataka', *Artibus Asiae*, 2005, 65: 259.

Chanchani, Aditi et al. (eds), *Sustainability in Tourism: A Rural Tourism Model*, Bangalore: Ministry of Tourism, Govt. of India/UNDP, 2008.

Coningham, R. A. E., 'Monks, Caves and Kings: A Reassessment of the Nature of Early Buddhism in Sri Lanka', *World Archaeology*, 1995, 27: 222–242.

Fontijn, David, 'The Significance of "Invisible" Places', *World Archaeology*, 2007, 39: 70–83.

Harrell, Mark, 'Sokkuram: Buddhist Monument and Political Statement in Korea', *World Archaeology*, 1995, 27: 318–335.

Heitzman, J., 'Early Buddhism, Trade and Empire', in K. A. R. Kennedy and G. L. Possehl (eds), *Studies in the Archaeology and Palaeoanthropology of South Asia*, New Delhi: Oxford University Press, 1984, pp. 121–137.

Murthy, A. V. Narasimha et al. (eds), *Excavations at Banavasi*, Mysore: Directorate of Archaeology and Museums, 1997.

Ray, Himanshu Prabha, *Monastery and Guild: Commerce Under the Satavahanas*, New Delhi: Oxford University Press, 1986.

Ray, Himanshu Prabha, 'The Shrine in Early Hinduism: The Changing Sacred Landscape', *The Journal of Hindu Studies*, 2009, 2: 76–96.

Ritti, S., 'Buddhism in Kannada Inscriptions', in Devendra Handa and Ashvini Agrawal (eds), *Ratna-Chandrika: Panorama of Oriental Studies (Shri R. C. Agrawala Festchrift)*, New Delhi: Harman Publishing House, 1989, p. 316.

Sastry, R. Sesha, 'The Memorial Stones of Karnataka', in D. V. Devaraj and C. B. Patil (eds), *Art and Architecture in Karnataka (Papers Presented at the National Seminar on Archaeology 1985)*, Mysore: Directorate of Archaeology and Museums, 1996, p. 125.

Settar, S., and Ravi Korisettar, 'Niśidis in Karnataka – a Survey', in S. Settar and G. D. Sontheimer (eds), *Memorial Stones: A Study of Their Origin, Significance and Variety*, Dharwad, New York and Heidelberg: Institute of Indian Art History, Karnatak University, The South Asia Institute and University of Heidelberg, 1982, pp. 283–287.

Shastri, A. M., 'Buddhism in the Deccan During the Sātavāhana Age', in Arundhati Banerji (ed.), *Hari Smriti: Studies on Art, Archaeology and Indology (Papers Presented in Memory of Dr. Haribishnu Sarkar)*, New Delhi: Kaveri Books, 2006, pp. 71–83.

Suvrathan, Uthara, 'Complexity on the Periphery: A Study of Regional Organization at Banavasi, C. 1st–18th Century A.D.', unpublished Ph.D. dissertation, University of Michigan, Ann Arbor, MI, 2013.

Suvrathan, Uthara, 'Regional Centres and Local Elite: Studying Peripheral Cores in Peninsular India', *Indian History*, 2014, 1: 89–142.

Vogel, J. P., 'No. 1 Prakrit Inscriptions From a Buddhist Site at Nagarjunakonda', *Epigraphia Indica*, 1929–30, XX: 7–8, 23.

Part II
Case studies of World Heritage Sites in India

5 Monumentality, nature and World Heritage monuments

The rock-cut sites of Ajanta, Ellora and Elephanta in Maharashtra

Pia Brancaccio

The inscription of sites in the UNESCO World Heritage List is a complex process tied to political and economic equilibria, as eloquently argued by Lynn Meskell.[1] In recent times, much discussion has highlighted the elusive boundaries surrounding the categories of cultural and natural heritage as applied to sites included in the World Heritage List. It has been noted that the duality implied by the juxtaposition of culture and nature is a product of Western processes, especially unfit to capture South Asian realities where profound and continued interactions exist between man and landscape.[2] The most recent addition to the World Heritage List in India, the Khangchendzonga National Park in Sikkim, reflects an increasing effort to bridge the categories of nature and culture within the UNESCO practice.[3] This is the first mixed site of cultural and natural relevance in South Asia, inscribed because of its biodiversity and landscape dotted with caves, rivers and lakes sacred to the local Buddhist tradition.

This chapter focuses on the rock-cut sites of Ajanta, Elephanta and Ellora in Maharashtra, India, inscribed decades ago in the World Heritage List due to their outstanding cultural relevance.[4] The elaborate architectural designs and visual ornamentations of these unique monuments carved on rocky cliffs still stand today as a testament to the impressive skills of unknown artists. UNESCO recognised the artistic value of these ancient Buddhist, Hindu and Jain sites while overlooking their deep religious and historical interconnectedness with the surrounding landscape. Re-evaluating epigraphic, architectural and artistic evidence extant at the caves, will demonstrate that the natural milieu was integral to the conception of these rock-cut temples, and that human activity at these sites was shaped by nature, rather than superimposed onto it. As we consider the monuments at Ajanta, Ellora and Elephanta only on the basis of their built environments, or better to say "carved environments", we obliterate their very essence and compromise their intended meanings.

When it comes to a site like Ajanta (which was cut in the ravine of the Waghora river between the 1st century BCE and the 5th century CE), or any other rock-cut World Heritage monument in Maharashtra, the deep interrelations between the natural environment and the man-made interventions are immediately apparent if not obvious: the siting of the monuments and the medium employed for the realisation of the *viharas* (monasteries) and *chaitya* halls speak for a complete integration of architecture, history and nature (Figure 5.1). Yet the relationship between the caves and their natural environments is not only physical and appears to be embedded into the conceptual fabric of the monuments. All evidence contemporary with the cave sites

Figure 5.1 View of the Ajanta caves.

Source: Courtesy of the author

indicates that these rock-cut structures were not conceived as simple mountain hermitages. Rather, they were intended to evoke palatial abodes in mythical paradises where the natural landscape, conjuring heavenly gardens, defined the parameters of this conceptual and visual metaphor. By re-establishing in historical terms the interconnectedness of human activity and nature at these cave sites, the inadequacy of their current classifications on the World Heritage List becomes apparent.

Caves have always played a key role in the Indian religious imaginary, and have been especially important in the practice of ascetic life across many religious traditions. The interaction between the practitioner and the untamed natural environment was seen as conducive to the attainment of a superior state of mind.[5] Within the Buddhist tradition, the importance of caves is illustrated in the well-known episode of the Indasalaguha told in the Pali *Sakkapanna suttanta*, where the Buddha himself engaged in ascetic exercises in the Indasala cave located near Rajagaha (Rajgir), surrounded by the forest and wild beasts.[6] The god Indra, also referred to as Sakka, visited him while in the cave to seek answers to 42 questions about the nature of existence. Indra sent ahead his divine musician, Pancasikha, and then personally met the Buddha, who extended his life by many cosmic cycles.

Depictions of the *Indasalaguha* story are frequently found in early Indian art: at Bharhut, in the 1st century BCE, the scene was represented on a *vedika* pillar that was re-cut at a later time (Figure 5.2).[7] In the image labelled "*Idasalaguha*", the wilderness and the environment are prominently emphasised: the scene is dominated by a natural, large *guha* or natural cave with a rocky floor, with vivid images of wild animals and jagged boulders above; some of the details of the representation are lost due to later reuse of the carving. Remarkably, in all later depictions of *Indasalaguha* throughout the 1st centuries CE, we witness a progressive transformation of the natural cave (*guha*) into a domesticated locale.

In the *Indasalaguha* episode depicted at the beginning of the Common Era on a pillar of the north *torana* (gateway) of the Sanchi stupa, the barren cave in the wilderness appears to be an elaborate rock-cut *chaitya* hall, obviously making reference to those cut in the basaltic rock of the western Deccan (Figure 5.3). At Sanchi, not only did the artists substitute an image of a natural cave with that of a man-made rock-cut *chaitya* hall, but they also transformed the surrounding landscape into a controlled natural environment. The mountainous setting is still suggested by the bulky blocks of stone, yet nature is represented as tamed and mannered: to the right of the *chaitya* hall we see two

Figure 5.2 Indasalaguha, detail of *vedika* pillar, Bharhut stupa.

Source: Acc. no. 68604, Neg. no. 483.34, American Institute of Indian Studies, Gurgaon

Figure 5.3 Indasalaguha, detail of *torana* pillar, Sanchi stupa.

Source: Courtesy of the author

mythical beasts crouching amidst rocks, and on the upper left side two tamed felines pose for the viewer. There is little evocative of a challenging wilderness here, and this shift from *guha* to *chaitya* hall, and from wilderness to garden, may invite us to reflect on the important role played by nature in association with these rock-cut monuments. As the Buddhist caves take the shape of elaborate architectural structures, the surrounding landscape is framed accordingly and loaded with new meanings and cultural connotations.

Let us begin by examining some of the rock-cut monuments contemporary with the Sanchi reliefs to shed light on the particular relationship between monuments and nature. By the 1st century BCE, when the Satavahana rulers controlled much of the Deccan, a large number of rock-cut monasteries were established in the hills of the Western Ghats. These man-made caves were always set in proximity of water resources and were immersed in a lush natural environment that included forests and animals. Buddhist sites located on the rugged basaltic cliffs of the Deccan plateau surprisingly did not include simple *guhas,* or natural rock shelters, but rather consisted of elaborate structures carved in the living rock that evoked urban milieus.

The living rock came to be a pliable medium in the hands of the capable stone cutters who created ornate residential units and palatial apsidal halls. At the site of Bhaja, one of the earliest in western Deccan established in the 1st century BCE, the *chaitya* hall displays a very elaborate and un-natural-looking façade topped by balconies and arches with overlooking couples (Figure 5.4). This rock-cut version of a multi-storied wealthy mansion, befitting an urban setting, seems to be completely out of context on a steep cliff surrounded by jungle. The monastic cells cut on either sides of the *chaitya* are integrated within this evocative choreography that contains no references whatsoever to forest hermitages. The façade of the *chaitya* complex at Bhaja is reminiscent of the cityscape captured in an episode of the life of the Buddha depicted on the east gate of the Sanchi stupa.[8] In the left part of the crossbeam illustrating the Great Departure, we can see elaborate buildings within the city walls that include mansions with balconies, arches, ladders and onlookers similar to the architectural ornamentations on the Bhaja façade.

Donors' inscriptions from the western Deccan caves may indicate how such imposing rock-cut structures and their locales were framed in contemporary views. A donor epigraph carved on a wooden rib of the *chaitya* hall at Bhaja clearly identifies the structure as a *pasada* (palace).[9] The same term *pasada* also appears in a roughly contemporary Buddhist inscription from the site of Bharhut, on a *torana* pillar known as the Ajatasatru pillar. It is carved on the dome of a building depicting a worship scene of the Buddha's headdress relic in the Assembly Hall of the Devas in the Trayastrimsa (Figure 5.5).[10]

The label "Vijayanta Pasada" identifies the structure as the legendary Vijayanta palace of the gods located in Trayastrimsa heaven. Not surprisingly, the Vijayanta palace looks very much like the *chaitya* hall in Bhaja with its horseshoe shaped portal, rounded roof, and

Figure 5.4 Karle *chaitya* hall.
Source: Courtesy of the author

tall, multistoried structure with balconies and alcoves from which on-looking devas emerge. Given the strong visual link between the Vijayanta palace from Bharhut and the palatial architecture from Bhaja, I suggest that the Bhaja's donors who referred to the *chaitya* as being a "*pasada*" intended to erect a structure that recreated the legendary deva's abode. The *chaitya* hall was conceptually associated with the palace of the god Indra where the relics of the Buddha were worshipped, and the couples represented on the Bhaja façade were intended to represent devas and not ordinary beings. By extension, the mountain on which the caves were cut became the mythical Mount Meru, with Trayastrimsa heaven located atop, and the natural surroundings of the caves came to be identified with the legendary gardens of the god Indra who presided over this particular heaven. The Trayastrimsa was the second heaven in Kamadhatu that

Figure 5.5 Worship of the Buddha's headdress in the Assembly Hall of the Devas, detail of *Torana* pillar, Bharhut stupa.

Source: Courtesy of the author

still maintained a physical connection with the rest of the world; it had a magical garden populated by trees that could heal, rejuvenate, enhance fertility or keep evil spirits at bay.[11]

Evidence from other early sites in the western Deccan confirms that rock-cut structures were intended to evoke lavish milieus associated with gods and heavenly gardens. The *chaitya* at Karle also seems to reproduce complex urban architectures and includes a very ornate porch and façade populated by life-size loving couples, or *mithunas*, while the interior is decorated by sumptuous pillars with pot-shaped capitals crowned by elegant animals and riders (Figure 5.6).

These architectural features are not just extravagant creations of the rock carvers from the western Deccan – in the reliefs carved on the east gate of the stupa at Sanchi we see similar elements (Figure 5.7). In Sanchi, elaborate pillars identical to those in the Karle *chaitya* hall occur in images of palaces worthy of the gods like the one we see here, where a royal figure surrounded by courtiers appears seated in *lalitasana* and protected by a parasol. Scholars have proposed that this image from Sanchi may in fact represent Trayastrimsa heaven.

An inscription from the *chaitya* hall at Karle confirms that donors consciously alluded to the *chaitya* as Indra's palace in Trayastrimsa heaven. A record inscribed in the porch mentions the establishment of a rock mansion, the most excellent in Jambudvipa, and makes reference to Vijayanti, perhaps not a town in Konkan as suggested in

Figure 5.6 Bhaja, *chaitya* hall.
Source: Courtesy of the author

readings of the inscription, but rather an allusion to the legendary palace of the Trayastrimsa heaven located at the epicentre of Jambud-vipa.[12] If these caves are replete with visual and epigraphic suggestions to heavenly palaces, then the natural environment shaping their very

Figure 5.7 Indra's heaven, detail of *torana* pillar, Sanchi stupa.

Source: Courtesy of the author

essence becomes the legendary garden of Indra's heaven described in texts as a beautiful place with magical trees, chattering monkeys and chirping birds.

Such literary topos dominate the epigraphic records from the Buddhist caves at Ajanta, which consist of 26 units cut in the basaltic rock of the Deccan plateau between the first BCE and the fifth CE (Figure 5.1).[13] The earliest core of the site around *chaitya* halls 9 and 10 was significantly expanded in the 5th century under the patronage of the Vakataka king Harishena and his entourage. At this later time two more *chaitya* halls (19 and 26) and 20 more *viharas* were carved in the living rock and embellished by elaborate paintings of scenes from the *jatakas* and *avadanas* or narratives related to the previous life of the Buddha, and other Buddhist images that survive only in part.

The Ajanta monastic complex was inscribed in the UNESCO World Heritage List (no. 242) in 1983 because it fulfilled the selection criteria nos. i, ii, iii and vi.[14] Ajanta's inclusion in the World Heritage List was based solely on the historical value of the human activity that transformed the exceptional ravine into a monastic site. Yet it is the siting of the Ajanta monastery in a lush paradise-like environment that defines the very essence of the architectural interventions, all suggestive of divine abodes. An inscription by the patron Buddhabadra in *chaitya* hall 26 describes the natural environment of the caves as a celestial garden. The epigraph reads: 'This cave has been established . . . on the top of the mountain which is frequented by great yogins, and the valleys of which are resonant with the chirping of the birds and the chattering of the monkeys'.[15]

Remarks made by Harishena's minister, Varahadeva, in verses 24–27 of the Cave 16 inscription at Ajanta also evoke the cave's architectural beauty in terms of Indra's paradise:

> [The dwelling] ornamented with windows, doors, splendid verandahs, railings and images of the Devakanyas and delightfully arranged pillars with *Chaitya*-Mandira. . . . A large reservoir of water and [adorned] by the abode of the chief of serpents and others. . . . Warmed in summer by the heat of the sun, and fit for enjoyment at all seasons . . . [as] the dwelling of Indra and the bright caves of Mandara . . . in the mountain to which none is equal in greatness . . . [he] made with love, pleasure and expansive modesty . . . a cave brilliant with the radiance of the crown of Indra.[16]

Here cave 16 is clearly equated to Indra's palace in paradise, and the landscape as a heavenly garden completes this literary allusion.

Note that the very terms *viharas* and *mandapas* occurring in Buddhist inscriptions to identify the caves are commonly found in Sanskrit poetic literature in association with courtly gardens.[17]

This kind of convergence of culture and nature that constitutes the conceptual framework of the Ajanta caves is also apparent at the cave monuments of Ellora and Elephanta in Maharashtra. Visual references to heavenly landscapes are recurrent at these two major rock-cut sites that reverberate (in/through?) the mythical geography of India. The cave temples of Ellora were inscribed in the World Heritage List (no. 243) at the same time as Ajanta, based on criteria i, ii and vi.[18] Ellora comprises 34 caves, including Hindu, Buddhist and Jain spanning from the 6th to the 10th centuries CE.[19] The first group of caves was established in the 6th century CE by Shaiva patrons and appears to be stylistically related to the Elephanta caves; in the 7th century the Buddhist occupied the southernmost edge of the cliff with caves 2 to 12, and in the 9th century the Jains created more elaborate caves (30–34) on the northern edge end of the site.

The best known cave at Ellora is undoubtedly the spectacular 8th-century CE Kailashnatha cave (no. 16) carved out of the living rock as a free-standing temple at the time of the Rashtrakuta king Krishna I (Figure 5.8). This rock-cut temple is covered by elaborate carvings and echoes the shape of Shiva's mountainous abode; it stands completely free from the rock cliff where surrounding multistoried pillared halls and rectangular chapels are excavated. In this particular case, the oneness of nature and the cultural intervention is evident and needs no explanation. The identity between the temple and the cosmic mountain Meru or Mandara is unmistakable and the monument appears to be *svayambhu*, naturally manifesting itself out of the living rock.

This perception is confirmed by an inscription. The well-known Rashtrakuta grant of Karka Suvarnavarsha dated to the 812 CE informs us of the identity of the patron of the Kailashnath cave, King Krishna I.[20] The grant overtly praises the temple dedicated to Shiva as *svayambhu*, and the patron's involvement in the creation of the monument is associated with the wondrous qualities of the gods. The inscription also explicitly compares the cave to a heavenly structure where the best of immortals move in celestial *vahanas* (mounts). Much like at Ajanta, the natural environment is not merely in the background at the site of Ellora's Kailashnatha cave but also shapes the very essence of the architectural intervention: it epitomises the siting of paradise where mansions of the gods are located.

Figure 5.8 View of the Kailashnatha from above, Ellora.
Source: Courtesy of the author

Nature defines the *raison d'être* of the cave monuments perhaps nowhere more than at Elephanta. The suggestive landscape of this island located off the coast of Mumbai is home to the beautifully sculpted Great Cave dedicated to the god Shiva at the beginning of the 6th century CE (Figures 5.9 and 5.10). Generally overlooked are the six minor Shaiva caves at Elephanta carved on the western and eastern slopes on the island and an early Buddhist stupa now in ruins near the top of the hill.[21] Elephanta was inscribed in the World Heritage List (no. 244) based on criteria i and iii, which refer uniquely to the superb sculptures of the Great Cave, in particular the so-called Sadashiva image, which refers to one of the identifications proposed.[22] The natural environment which historically empowered the cave site is completely overlooked in the UNESCO report. Yet it was its uniqueness that endowed the Great Cave with "cosmic relevance": the island in the Indian Ocean carries clear allusions to Jambudvipa, the Rose Apple island surrounded by concentric oceans of salt, and beyond to the whole universe identified in Hindu cosmology as a closed egg floating on cosmic water. In essence, the power of this monument lies

Figure 5.9 Elephanta, view of the harbour.

Source: Courtesy of the author

Figure 5.10 So-called Sadashiva, Great Cave, Elephanta.
Source: Courtesy of the author

almost completely in its natural surroundings, and much less in the ability of the skilled artists who crafted the wondrous images of Shiva in the Great Cave.

This makes apparent how natural elements, as much as historic interventions, define the essence of the rock-cut World Heritage Sites of Ajanta, Ellora and Elephanta. Through a re-examination of the epigraphic and architectural evidence available at several rock-cut sites, it is possible to retrace the original meaning attached by patrons to cave temples and their natural environments. Nature empowered these monuments in the religious imaginations of the devotees, transforming the Buddhist and Hindu caves into pavilions within divine gardens. This integral relationship between monasteries, temples and their surroundings begs us to reconsider the absolute value placed on the distinction between cultural and natural heritage when considering the rock-cut monuments of Maharashtra. It is almost as if, while assessing the heritage value of the Taj Mahal, one would completely overlook its surrounding gardens therefore obliterating the very meaning of the monument intended as a visual metaphor of paradise.

Notes

1 Lynn Meskell, 'UNESCO's World Heritage Convention at 40: Challenging the Economic and Political Order of International Heritage Conservation', *Current Anthropology*, 2013, 54(4): 483–494.
2 Denis Byrne and Gro Birgit Ween, 'Bridging Cultural and Natural Heritage', in Lynn Meskell (ed.), *Global Heritage: A Reader*, Somerset: Wiley Blackwell, 2015, pp. 94–111.
3 UNESCO World Heritage Center, 'Khangchendzonga National Park', http://whc.unesco.org/en/list/242 (accessed on 30 April 2017).
4 Ajanta and Ellora were inscribed in the UNESCO World Heritage List in 1983 while Elephanta was inscribed in 1987. UNESCO World Heritage Center, 'Ajanta Caves', http://whc.unesco.org/en/list/242; 'Ellora Caves', http://whc.unesco.org/en/list/243; 'Elephanta Caves', http://whc.unesco.org/en/list/244, http://whc.unesco.org/en/criteria/ (accessed on 30 April 2017).
5 Phyllis Granoff, 'What's in a Name? Rethinking "Caves"', in Pia Brancaccio (ed.), *Living Rock: Buddhist, Hindu and Jain Cave Temples in the Western Deccan*, Mumbai: Marg, 2013, pp. 18–29.
6 Thomas W. Rhys Davids, *The Digha Nikaya*, Vol. II, London: Pali Text Society, 1903, pp. 263–275.
7 Alexander Cunningham, *The Stupa of Bharhut*, London: W. H. Allen & Co., 1879, p. 128 no. 92, plate XXVIII.
8 Vidya Dehejia, *Indian Art*, London: Phaidon, 1997, pp. 52–53, Figure 34.
9 S. Nagaraju, *Buddhist Architecture of Western India*, New Delhi: Agam Kala Prakashan, 1981, p. 329, no. 3.
10 Alexander Cunningham, *The Stupa at Bharhut*, London: WH Allen & Co., 1879, p. 137 no. 65, plate XVI.
11 Akira Sakadata, *Buddhist Cosmology: Philosophy and Origins*, Tokyo: Koei, 1997, pp. 25–67.
12 James Burgess and Indraji Bhagwanlal, *Inscriptions From the Cave-Temples of Western Deccan*, Bombay: Governemnt Central Press, 1881, p. 28, no. 1.
13 For a comprehensive history of Ajanta see Walter Spink, *Ajanta: History and Development*, Vols. 1–5, Leiden: EJ Brill, 2005–2014.
14 UNESCO World Heritage Center, 'Ajanta Caves'. The inclusion of the Ajanta caves in the World Heritage List was motivated by the following selection criteria: i. to represent a masterpiece of human creative genius; ii. to exhibit an important interchange of human values, over a span of time or within a cultural area of the world, on developments in architecture and technology, monumental arts, town-planning or landscape design; iii. to bear a unique testimony or at least exceptional testimony to a cultural tradition or to a civilization which has disappeared; vi. to be directly associated with ideas, beliefs and artistic works of outstanding universal significance. UNESCO World Heritage Center, 'The Criteria for Selection', http://whc.unesco.org/en/criteria/ (accessed on 30 April 2017).
15 I use here the rather poetic translation of the Ajanta Cave 26 inscription published by Spink, *Ajanta: History and Development*, Vol. 1, p. 420.

16 Burgess and Bhagwanlal, *Inscriptions From the Cave-Temples of Western Deccan*, p. 72.
17 Daud Ali, 'Gardens in Early Indian Court Life', *Studies in History*, 2003, 19(2): 221–252.
18 UNESCO World Heritage Center, 'Ellora Caves' and 'The Criteria for Selection'.
19 On the various phases of occupation at Ellora, see M. K. Dhavalikar, *Ellora*, New Delhi: Oxford University Press, 2005; Geri Malandra, *Unfolding a Mandala, the Buddhist Caves at Ellora*, Albany, NY: State University of New York Press, 1993; Lisa Owen, *Carving Devotion in the Jain Caves at Ellora*, Leiden: EJ Brill, 2012.
20 R. G. Bhandarkar, 'The Baroda Copper Plate', in N. B.Utgikar (ed.), *The Collected Works of Sir R.G. Bhandarkar*, Vol. III, Pune: Bhandarkar Oriental Institute, 1927.
21 For a scholarly overview of the site of Elephanta, see George Mitchell, *Elephanta*, Mumbai: India Book House, 2002.
22 UNESCO World Heritage Center, 'Elephanta Caves'.

References

Ali, Daud, 'Gardens in Early Indian Court Life', *Studies in History*, 2003, 19(2): 221–252.

Bhandarkar, R. G., 'The Baroda Copper Plate', in N. B. Utgikar (ed.), *The Collected Works of Sir R.G. Bhandarkar, Vol. III*, Pune: Bhandarkar Oriental Institute, 1927.

Burgess, James, and Bhagwanlal Indraji, *Inscriptions From the Cave-Temples of Western Deccan*, Bombay: Government Central Press, 1881.

Byrne, Denis, and Gro Birgit Ween, 'Bridging Cultural and Natural Heritage', in Lynn Meskell (ed.), *Global Heritage: A Reader*, Somerset: Wiley Blackwell, 2015, pp. 94–111.

Cunningham, Alexander, *The Stupa of Bharhut*, London: WH Allen & Co., 1879.

Dehejia, Vidya, *Indian Art*, London: Phaidon, 1997, pp. 52–53, Figure 34.

Dhavalikar, M. K., *Ellora*, New Delhi: Oxford University Press, 2005.

Granoff, Phyllis, 'What's in a Name? Rethinking "Caves"', in Pia Brancaccio (ed.), *Living Rock: Buddhist, Hindu and Jain Cave Temples in the Western Deccan*, Mumbai: Marg, 2013, pp. 18–29.

Malandra, Geri, *Unfolding a Mandala, the Buddhist Caves at Ellora*, Albany, NY: State University of New York Press, 1993.

Meskell, Lynn, 'UNESCO's World Heritage Convention at 40: Challenging the Economic and Political Order of International Heritage Conservation', *Current Anthropology*, 2013, 54(4): 483–494.

Mitchell, George, *Elephanta*, Mumbai: India Book House, 2002.

Nagaraju, S., *Buddhist Architecture of Western India*, New Delhi: Agam Kala Prakashan, 1981.

Owen, Lisa, *Carving Devotion in the Jain Caves at Ellora*, Leiden: EJ Brill, 2012.

Rhys Davids, Thomas W., *The Digha Nikaya*, Vol. II, London: Pali Text Society, 1903, pp. 263–275.

Sakadata, Akira, *Buddhist Cosmology: Philosophy and Origins*, Tokyo: Koei, 1997.

Spink, Walter, *Ajanta: History and Development*, Vols. 1–5, Leiden: EJ Brill, 2005–2014.

6 Removable heritage
Nalanda beyond the *Mahavihara*

Salila Kulshreshtha

Introduction

UNESCO's listing of Nalanda as a World Heritage Site identifies its "Outstanding Universal Value" (OUV) as a site which 'comprises the archaeological remains of a monastic and scholastic institution dating from the 3rd century BCE to the 13th century CE. It includes stupas, shrines, viharas (residential and educational buildings) and important art works in stucco, stone and metal'.[1] My chapter attempts to analyse the spatial and temporal parameters within which the heritage zone of Nalanda has been defined and how an extension beyond the currently excavated boundaries to embrace broader aspects of the geographical and sacred landscape would extend the implication of Nalanda beyond its scope as a Buddhist *Mahavihara*.

I shall examine how such state-directed construction of history and subsequent delineation as "heritage" echo the 19th and 20th century colonial paradigm which makes monuments out of religious sites, devoid of their archaeological and social contexts. Following frameworks drawn up in the colonial period, places of habitation and worship are studied in light of textual accounts or as architectural "wonders" devoid of living traditions, participation in ritual networks or as placed within cultural narratives. Nalanda has been taken as the point of origin of pedagogy, philosophy, religion, art and architectural design. Its impact all over Asia has been highlighted but its interaction with the immediate hinterland stands neglected. The site has been conceptualised merely in terms of its Buddhist remains with the association of the different *chaityas* (a Buddhist shrine or prayer hall with a stupa at one end) and stupas with the life of the Buddha and his disciples. In such an endeavour the topographies of other religions have been marginalised and the *Mahavihara* has been looked upon as a self-contained unit without much interaction with the immediate hinterland. The focus on the excavated area – its planning, purpose and orientation – has failed to accommodate the context of sculptures and other religious icons found at the site.

Is it possible to view a large monastic complex, such as at Nalanda, as an isolated compound with no cultural exchange with settlements in its close vicinity? I will examine the religious topography in the immediate hinterland of the excavated area, to draw out the inter-mixing of faiths and ritual praxis and to emphasise upon a sacred architectural complex around the *Mahavihara*. My chapter traces the evolution of Nalanda as a sacred sub-region post 9th to 10th centuries CE, taking into account the history of shrines and sacred iconography surviving from villages of Begumpur, Surajpur, Bargaon, Jagdishpur and Kundalipur in the vicinity of the present archaeological site.

The extensive remains of sculptures of Hindu deities and shrines from the villages indicate a simultaneous coexistence of Shaiva, Vaish-nava and Surya sects in the region along with Buddhism. A large cor-pus of Hindu and Jain sculptures and sacred motifs from Nalanda also survive in various museums and private collections. Aerial and satellite images in recent years have redefined the boundaries of Nal-anda beyond the present-day excavated area. My study will take into account sites, structures and sculptures to highlight a larger religious complex where sacred space was shared by Buddhists, Jains and Hin-dus, where mythologies and rituals came to be closely interwoven and a monotheistic affiliation of the site becomes difficult to imagine.

Early encounters

I will begin by tracing the various colonial narratives on the discovery of Nalanda and how in the course of the search for a Buddhist topogra-phy evidence of remains belonging to other religions became incidental. Since the time of its earliest visitor a multicultural character is discern-ible for the larger catchment area surrounding the monastic complex.

Francis Buchanan arrived at Nalanda in 1812 and recorded in his journal that about four miles after crossing the Panchana river, he came to a tank called "Dighi", from where the "ruins" commenced. Even at the time, he observed two separate groups of ruins and says that the southern group was larger of the two.[2] Immediately west of the Dighi *pokhar* (water tank) he found an elevated mound with frag-ments of bricks scattered about; the 'north end is occupied by part of the village Begumpur, and a small ruinous mud fort erected by Kamgar Khan'.[3] Buchanan also mentions seeing four separate mounds south of the fort and that on the eastern-most mound he finds two Jain images.[4] He also noted a web of tanks surrounding the site.

Buchanan came across heaps of Buddhist images which reminded him of Bodh Gaya and writes that many of these had been used in modern temples. He refers to the conical heaps of brick towards the south of Bargaon, suspecting them to be the remnants of temples. The living memory of the site, however, was not Buddhist but Hindu: he was told by the villagers that the ruins belonged to the ancient city of Kundalipur, famous in the epics and the *Puranas* as the capital of King Bhimaka of Vidarbha, the father of Rukmini, wife of Krishna. Buchanan therefore described the ruins accordingly as those of Kundalipur and the part of the site excavated afterwards represented, according to him, the ruins of a palace which 'consisted of various parts, the abode of the chief courtiers and officers of government. Among these may be traced some temples rising in conical mounds'.[5] The Jains who lived there, however, believed that the ruins represented the city of Pompapuri, which belonged to Raja Srenika, whom Buchanan identified as the ruler of Magadha.[6]

Buchanan also records contemporary religious practices such as an annual fair around the Suraj *pokhar* at Bargaon. In Bargaon, Buchanan also mentions seeing a Buddha image worshipped as *Bathuk Bhairava,* a form of Shiva, and the Rukministhan at Jagdishpur; both of these sites will be further discussed in the course of this chapter (Figure 6.1).

Figure 6.1 Buddha image at Rukministhan, Jagdishpur.

Source: Courtesy of the author

At the village of Kapatya, south of Bargaon, Buchanan mentions the temple of Kapatesvari, with many images collected around the main image of Buddhist divinity worshipped as goddess Kapatesvari, after whom the village was named.

Alexander Cunningham visited the site in 1861 and documented the spatial and topographical details of the site, mounds, water bodies and settlements almost to scale in relation to the surrounding topography. He prepared a detailed diagram which he titled, 'Sketch of the Ruins of Nalanda Mahavihara'.[7] Cunningham was the first to establish the identity of the site as Nalanda on the basis of its distance from Rajgir, but more so on account of two inscriptions which he found at the site mentioning its name as Nalanda. The first is a two-line inscription found on the back of an image of Vageshwari at Kapatiya, dated to the first year of Gopala Deva.[8] The second inscription is found on the back of the image of Panchika (Cunningham called this image 'Adi Shakti') and is dated to the third year of the reign of Devapala.[9]

In his report of 1861–1862 Cunningham writes that on reaching Bargaon, numerous masses of brick ruins were visible, which were probably the remains of gigantic temples attached to the famous monastery, though much of the remains were covered by square patches of cultivation.[10] Living memory attributed the remains as the ruins of the palace of one Raja Srenika and other legends simultaneously called these monuments the birthplace of Rukmini.[11] Cunningham, however, doubted the validity of these legends and was certain that he could prove 'that the remains at Bargaon are the ruins of Nalanda, the most famous seat of Buddhist learning in all India'.[12]

Cunningham visited the tree shrine at Jagdishpur village located about one and half kilometres southwest of the Nalanda monastery where under a huge neem tree several statues were collected, the most noteworthy being the monolithic sculpture depicting the Buddha in *bhumisparsamudra* ("Touching the earth" gesture of the Buddha), seated under the Bodhi Tree, surrounded by demons probably trying to distract him.[13] Cunningham noted that this statue was in fine state of preservation: it was enshrined and washed with milk and oil and covered with lead paint every day; the "ignorant" villagers worshipped this as an image of Rukmini. The ground was still wet with the blood of a recently killed goat.

A second sculpture that attracted his attention was at the entrance gate to Monastery 11 where a colossal stone image of the Buddha seated in the *bhumisparsamudra* was known and worshipped by the name of Bhairava, or Shiva in his terrible form, which he believed was the original image of the temple of Baladitya. The ground surrounding the image was called "Bhairo-sthan" or "Baithak Bhairav".[14]

At a distance of 2,000 feet to the north of the monastery, and to the east of the Suraj *pokhar*, Cunningham also came across the ruins of a very large temple. Owing to its close proximity to the village of Bargaon, this mound had supplied materials for all the existing houses (Figure 6.2).[15] He mentions a Hindu temple in the village and the house of the Zamindar Mitrajit where hoards of ancient images were stored and worshipped. He also records seeing in the village a four-armed Vishnu on Garuda. He visited the small temple in the hamlet of Kapatiya, where there were several "interesting" figures, amongst which were a Vajra Varahi and the inscribed Vageswari mentioned earlier, the latter carrying an inscription in two lines, which gives the name of the place as Nalanda, and is dated to the first year of the reign of the Pala ruler Gopala Deva.[16]

Cunningham also visited a Jain temple near Kamgar Khan's fort and noted that architecturally it resembled the temple at Buddha Gaya. He noted that it would be of about the same age but had undergone a modern renovation. The temple was white-washed and contained

Figure 6.2 Fragments of sculptures on the banks of the Suraj *pokhar*.
Source: Courtesy of the author

several Jain figures; the figure of Mahavir bore the date 1504 *Samvat,* or 1447 CE. He also notes that on the banks of the Suraj Kund many interesting figures were collected, mostly Buddhist, but there were also some figures of Vishnu, Shiva and Parvati, and Surya. When A. M. Broadley, the District Magistrate of Bihar Sharif, visited Nalanda in 1871 to explore the sites, he collected sculptures and antiquities which he displayed in a museum he established in his official premises in Bihar Sharif in 1878. Based on his study of sites and the antiquities, he concluded that Rajgir and Dapthu were "most ancient" while Bargaon was relatively new,[17] though he commented that the carvings at Rajgir were inferior to those found from Bargaon.[18] He also believed that Nalanda and Rajgir were the two most important Buddhist sites in Bihar.[19] In his detailed account of Bargaon he described high conical topes around the several tanks located there and identified them as remains of temples.[20]

On reaching Bargaon, Broadley noted two distinct mounds, the northern being the ruins of a Muhammedan fort, having no connection with the 'Buddhist remains'.[21] On visiting the northern mound he saw the ruins of two small stupas, some mutilated Buddhist statues and a fine figure of the Buddha. He also recovered several Hindu idols, including a crowned Vishnu seated on a 'sacred bird'.[22] To the southwest of the topes he saw the Suraj *pokhar,* flanked by ghats and a 'row of small pagodas on the north side, covered by massive brick cupolas, and their ruins still exist in tolerable entirety'.[23]

He mentioned seeing images of a booted Surya and another of Vishnu at Bargaon (Figure 6.3) and was able to draw parallels with similar images from elsewhere which he had in his collection. He also came across a second mound to the east with heaps of Buddhist carvings, door lintels and other antiquities. It was on this mound that he mentioned seeing a large Buddha being worshipped as Teliya Baba for whom he gives detailed measurements.[24] In 1871, Broadley began excavations on this main mound with 1,000 labourers. Within ten days he laid bare the eastern, western and southern façades of the great temple and published a short note of the excavation.[25]

Broadley visited the temple of Kapateshwari on the same mound and records how the mound was being gradually denuded because of brick robbery.[26] Broadley also explored the mound of Jagdishpur, where he came across the gigantic *alto-relievo* (high relief) figure of Buddha worshipped as the goddess Rukmini.

Based of the accounts of Chinese travellers, supplemented by survey and exploration of the site, Cunningham established the archaeological remains at Nalanda as those of a Buddhist university. He started

Figure 6.3 Inscribed, damaged Vishnu in a roadside tree shrine at Bargaon.
Source: Courtesy of the author

his excavation of the site in 1863, which was followed by Broadley's haphazard excavation in 1871 and a second spate of excavation by Cunningham commenced in 1872. The ASI purchased the site in 1916 and over the next two decades (between 1916 and 1937) several seasons of excavation were undertaken at the site along with attempts at preservation and collection of antiquities. Given the vast expanse of the remains and the limited funds, the ASI was able to purchase only a small tract of land. Excavation remained confined to this area, which was identified as that of a *Mahavihara*. Much of the area in the immediate vicinity of this "monastic complex" still remains unexplored. The site came to be listed as a Protected Site under the Ancient Monuments Preservations Act in 1904. The official discourse on the site framed by the ASI continues to look at the site through the prism of the colonial archaeologists – as a Buddhist *Mahavihara* – while the identification of the structures of the site are as per the narratives provided by the Chinese travellers.

The focus on *in situ* conservation in the 20th century provided the necessary impetus for the setting up of a site museum by the ASI in 1917 essentially to house the antiquities excavated from Rajgir,

Nalanda and other sites in the vicinity.[27] The museum currently has more than 13,000 antiquities in its possession, of which only 350 are displayed in its four galleries. These are dated between the 5th and the 12th centuries CE and have been collected from the ruins of the excavated area, from the neighbouring villages, from Rajgir and other sites in the vicinity. The thrust of the museum is to highlight the Buddhist antecedents of Nalanda, although it has a substantial portion of Hindu and Jain images in its collection.

Gallery number 1 houses Hindu, Buddhist and Jain sculptures. Of the four display cases three have Buddhist images and one is dedicated to the Hindu deities. Gallery number 2 displays miscellaneous objects unearthed during excavations in the monastic complex, including seals and sealings, ornaments, terracotta, stucco figurines, tools and implements and other objects of everyday use. Gallery number 3 displays Nalanda bronzes, including sacred images, while Gallery number 4 includes inscriptions and sculptures belonging to the "Pala Period" and concentrates on "Buddhist" antiquities.[28]

The current display, labelling and cataloguing of artefacts still reflects colonial taxonomies and the labels provide basic information giving the name, approximate date, material and location of find: 'There is no sense of Nalanda as a living space in which sculptures performed a function'.[29] Despite their display in a site museum the antiquities show no connection with the actual site in terms of architectural location giving details of where they were found. Neither does the signage in the monastic complex give examples of antiquities found there and now present in the museum. The display and arrangement at the museum makes a clear segmentation between the Buddhist and Hindu deities. Even though the icons have been collected from various sites, the narrative of the museum focuses only on the excavated ASI site.

The original scope of the ASI and the thrust of the museum on establishing the Buddhist identity of the site specially through the sacred images and motifs is well established through A. Ghosh's *A Guide to Nalanda*. In this publication the focus of the museum is on the Buddhist images and a very small fraction of Hindu images are listed.[30] The images and other antiquities from Nalanda are organised under dynastic labels with the "Gupta Period" as the beginning and "Pala Period" as a phase of last artistic interlude before the final collapse of the site due to Muslim invasions.[31] The aspects of "Tantric" religion and Nalanda as crucial in its development is highlighted. Religion and art from Nalanda are referred to as being points of contact and export of artistic and religious influences to Southeast Asia and to Tibet. The colonial categorisation of the Buddhist and Hindu faith in conflict with each other are made obvious by the use of sculptural examples such as the *Trilokavijaya*.[32]

Having discussed the discovery, documentation and early attempts at excavations in the monastic area, in the next section of the chapter I will map the hinterlands beyond the excavated area to identify the sacred structures and enclosures. This will be followed by plotting the various sacred networks across the different points on the map, which will highlight how the present-day excavated area of the ASI is a very small portion of the archaeological remains in Nalanda and is used to project a Buddhist heritage of the site.

Sacred mapping

An understanding of the possible extent of the site is apparent in the writings of the colonial travellers and explorers which suggest that the archaeological remains of the site are much larger than the present excavated area. Buchanan wrote that the ruins commenced from the Dighi *pokhar*. He records the nature of remains when he visited Bargaon, Kapatiya and Jagdishpur. Broadley approached the area probably from the north, from near Begumpur, and mentions Kamgar Khan's fort ruins and the area as quite scenic and surrounded by groves of mango trees.[33] He also explored the banks of the Suraj *pokhar*, Kaptiya and Jagdishpur.

The first attempt at mapping and understanding of the spatial remains of the site probably comes from Cunningham's 'Sketch of the Ruins of Nalanda Mahavihara', which was published in 1871.[34] Cunningham visited the villages mentioned by Buchanan and leaves behind his descriptions of mounds, tanks and shrines in the vicinity, giving a sense of the actual spread of the site and drawing various points of correlations in the historical geography of the site. From the rough sketch plan given by Cunningham the site appears to cover an area of about 3 kilometres in length and a mile in width east to west and shows the spatial distribution of villages, tanks and archaeological remains such as mounds, stupas, monasteries, temples, walls and sculptures.

Two studies in recent years based on the application of scientific technology have remarkably added to our understanding of the spatial layout and gradients of the site. A Cartosat study by J. Kamini and colleagues, undertaken in 2007, shows the extent of the site, patterns of settlement, shapes of excavated structures and the overall topography of the hinterland.[35] Based on the varied gradient of the area, the study suggests mounds which are so far unexcavated such as a possible stupa near Jagdishpur. Four tanks located at the cardinal points are also an indication of the perimeter of the site. It has been argued that the presumed "university" was spread over an area of 16 kilometres, out of which only 1.5 kilometres have been excavated so far.

A second study, undertaken by M. B. Rajani, was based on the application of remote sensing and Geographic Information Systems (GIS) to understand the expanse of the archaeological remains at Nalanda.[36] Rajani has studied the water bodies in the immediate vicinity of Nalanda to suggest a probable extent of archaeological remains at the site and has identified two large clusters of mounds (northern and southern) within these probable bounds. Based on these, she has argued that though south Bihar has several scattered water bodies, the ones clustered around Nalanda form a pattern not seen elsewhere in the vicinity. This suggests that the water bodies were possibly man-made and associated with the site, and their patterns of distribution may help trace its extent. She moreover argues that the tanks should not be studied in isolation but as interrelated, to draw out patterns of economic sustenance of the villages and the monastic complex. Rajani's GIS mapping has also revealed other boundaries for a site which extend much beyond the excavated area – such as fortification walls, moats and boundaries of neighbouring villages – indicating the spatial spread of the establishment rather than a definite boundary. Two satellite images show the two distinct mounds and the present-day villages that they contain, along with evidence of remains of two temples. Based on her imaging, Rajani has suggested the possible existence of a central axis joining the northern mound to the southern. This is apparent in the present-day organisation of the site by the ASI along a central path which divides the monasteries or residences on the west from the *chaityas* in the east. Looking at the larger picture at Nalanda, Rajani argues that though the name refers to the famous Buddhist monastery, the region is rich with Jain and Hindu antiquities as well.

My methodological approach takes into purview the satellite images of Nalanda produced by Rajani along with Google satellite images and the old-fashioned topography sheet of the area. By discussing the topography and spatial layout of the site, I will substantiate my methodology of looking beyond the excavated area to highlight several significant landmarks which have been sidelined: (a) the tanks and the settlement patterns; (b) the remains at *Serai Mound* or what has been called Temple 2, located just outside of the monastic complex; and (c) nature of remains in the surrounding villages of Bargaon, Begumpur,

Jagdishpur and Kapatiya. The site can be understood as comprised of two mounds. The southern mound is larger and includes the whole of the excavated area and adjacent regions, including the villages of Muzaffarpur, Kapatiya, Surajpur and Bargaon. The northern mound is smaller and covers Begumpur and its environs.

Tanks

The many *pokhras* in the vicinity of Nalanda were noticed and explicitly mentioned by Chinese travellers and British explorers. Both Buchanan and Broadley mention approaching the site from the environs of Dighi *pokhar*.[37] Cunningham also emphasises on the conspicuous presence of the 'noble tanks which surround the ruins on all sides'.[38] He recorded, 'To the North East are the Dighi *pokhar* and the Pansokhar *pokhar*, each nearly a mile in length; while to the south there is the Indra *pokhar*, which is nearly half a mile in length'.[39] To the south of the monastic complex he noticed a small tank, which was called Kargidya *pokhar* when he visited the site. Cunningham presumes that this was the Nalanda tank mentioned in Chinese accounts 'a tank in which the dragon, or *Naga Nalanda*, was said to dwell, and the place was named after him Nalanda'.[40]

The largest tank in the area is the Dighi *pokhar;* other major tanks include the Pansokhar *pokhar*, Indira *pokhar* and Suraj *pokhar*. Immediately south of the monastery are Rahela *pokhar* to the west and Kirgidya *pokhar* to the east. More recently B. N. Misra, in his monograph *Nalanda,* has listed the names of about 29 tanks and reported that local tradition actually speaks of 52 such tanks.[41] Others have suggested that the tanks were man-made and were excavated by removing mud to make bricks for the monastic area.[42] Rajani agrees that the tanks were man-made, however, based on the shape and distribution of the tanks she has argued that the 'tanks show careful planning: they are mostly geometrical (squares or rectangles), with sides roughly parallel to the four cardinal directions. Such precision would have been unnecessary if these tanks were excavated solely for purpose of mining earth for brick making'.[43]

The tanks are significant to the present study as their location shows the extent of the site and archaeological remains, and gives an idea of the general layout of the site, which spills much beyond the area excavated by the ASI. I will also draw upon present-day ritual and cult practices to trace sacred and mythological networks between the tanks, shrines and residential quarters.

Serai Mound

The second interesting site in the area is called "Serai Mound" and is believed to be the remains of a Hindu temple, also called Temple 2 or "Patthar Ghatti". Two features make this temple distinct from the planning of the rest of the monastic complex: unlike the brick built structures in the monastic complex, this temple is made of stone and lies at an angle from the perfectly aligned buildings of the monastery. The *jagati* (plinth) of the temple is decorated with 211 sculptured panels that include a range of themes such as scenes from Hindu mythology, images of Rama and Krishna, Shiva and Parvati, Kartikeya, Surya, Gaja Lakshmi, Kubera, etcetera. Except for the two panels containing scenes from the *jatakas*, Buddhist sculptures are conspicuously absent. It has been suggested that the temple might have been dedicated to a Hindu deity. It has also been argued that facing this temple site on the east is a lowland area, now under cultivation, which may well once have been a tank, which is normally an essential accessory to a sun temple.[44] Excavations have suggested that the present temple was built over a pre-existing brick structure and probably meant to replace it. On the basis of graffiti marks or pilgrims' records on the northeast part of the *jagati* the addition of the stone edifice to the temple has been assigned to the 6th or 7th centuries CE.[45]

Surajpur and Bargaon villages

The Suraj *pokhar* is situated to the northwest of the excavated area and is marked by two large settlements, Surajpur and Bargaon, which almost merge into each other. Broadley wrote:

> The tank called Suraj Pokhra about 400 feet square, was once flanked with a row of small topes on the north side covered with massive brick cupolas, and their ruins still exist in tolerable entirety. I clearly marked six of these temples. On each side of the tank were three brick ghats and the ruins of these may still be traced. The bank of the tank served also as the repository of the chaityas. Several of these were taken out of the tank by me and I saw many others beneath the clear water. At the South East corner of the pond I found a perfect heap of idols, all of great beauty; and the receding waters had laid bare an enormous and elaborately carved Varaha, 9 feet high and 4 feet wide, broken in two pieces. Most of these are now in my collection.[46]

Some of these – ruins particularly those around the sacred tank – had also earlier attracted the notice of Buchanan. The two villages are dotted with shrines in which older images have been reinstated and the shrines given brand-new names to show adherence to the different sects of the Hindu pantheon. The houses in the village show reuse of architectural fragments from older structures as door steps, lintels and pillars.

Begumpur

Situated at the western margin of the Dighi *pokhar* is the village of Begumpur. About 400 feet to the south of the hamlet is a large square mound with a ruined mud fort, which according to Buchanan was built by Kamdar Khan, a military adventurer of the 18th century. Immediately to the south of this, in 1871 Broadley had noticed two Buddhist stupas with Buddhist and Hindu idols including the image of Vishnu on Garuda. Cunningham had also made note of the two topes, from where he also recovered two Jain images, and identified these as the site of the current Jain temple. Begumpur probably marks the northern extremity of the site.

Rajani's satellite images of the mound supplemented with surface explorations has revealed the evidence of a brick structure under the mound. She further suggests that the Buddhist establishment at Nalanda included the northern mound at Begumpur where an additional temple might have been located and was joined with the temples in the southern mound through a north–south axis. In a rather far-fetched theory derived from the accounts of the Chinese travellers, she also suggests that this complex at Begumpur which has so far remained unexcavated might be the lost monastery of Oddantapuri, which is believed to have been located a short distance away from Nalanda.

Jagdishpur

Located about 2 kilometres away from the excavated site is a 200 feet high square mound on top of which is a single-celled shrine with a towering *shikhara*. The site was visited by Buchanan, Broadley and Cunningham. Cunningham mentions a colossal, 15 feet high and 9.5 feet wide Buddha in *bhumisparsamudra* here as the "object of worship". When Cunningham had visited the site it had no built structure but the colossal Buddha along with other sculptures were preserved and worshipped under a tree.

Kapatya

The village apparently derived its name from the ruins of the temple of goddess Kapatesvari noticed by Buchanan as having an enshrined image of the goddess representing a "fat male".[47] It is not clear what this image really represented, but Buchanan was sure that the numerous images collected around the shrine were mostly those of Buddhist deities, on some of which he observed inscriptions. Cunningham also noticed the temple with a collection of several interesting figures including a fine Vajra Varahi and an inscribed Vageswari. No such temple and collection of sculptures are now to be seen at the village as the images are now in the Indian Museum.

Sacred networks

One way of looking at the site has been to trace its vertical connections – how the sites, shrines and sculptures have evolved over the centuries – which has been the most popular trend of studying the site so far.[48] The surviving structures at the monastic complex, stray icons and religious motifs are presented under different chronological periods, which break the continuity of habitation at the site: the periods become stratified compartments. I propose here to highlight the horizontal linkages between different points of the settlements to argue that Nalanda did not receive its final deathblow in the 12th–13th centuries but lived on in the cultural memory of the people with several reinvented identities.

From the 13th century onwards, Nalanda the monastery seems to have gone out of existence and oral traditions began to relate to the extensive ruins – consisting mainly of earthen mounds strewn with bricks and carvings – as representing the site of the ancient town of Kundalipur. The Hindus of the locality started appropriating some of the carvings and images from the ruins for their own worship as Hindu divinities. Bargaon also finds mention as Vatagram or Baragram in the Jain chronicles of the 16th and 17th centuries.

I begin my discussion with the southern mound, the area immediately north of the excavated site; based on my field study in these villages. I will trace the relationship between tanks and shrines; the tanks are surrounded by ghats around which a series of modern shrines are located with no fixed iconographic programmes. The shrines show a haphazard collection of ancient sculptures and architectural fragments; in most cases the latter are also worshipped and duly anointed. Often the legends around the shrines are closely interwoven with a sacred tree: the pipal, banyan or neem. Through my discussion I will sketch a ritual network between the shrine, the tanks and the trees.

Bargaon and Surajpur around Suraj pokhar

There are two principal sacred landmarks around the Suraj *pokhar*: the tank itself, which is integrated into many oral legends and the Surya *mandir* (temple dedicated to Sun god) in the heart of the village. The Suraj *pokhar* is surrounded by ghats on three sides. Several modern shrines with a rich collection of ancient sculptures and architectural fragments dot the banks. Clearly images from elsewhere have been reinstated in these temples. The images from Bargaon are mostly Hindu, typically the black basalt images popular in East India and the images are unusually large in size, some over 4 feet high. An almost life-size inscribed Vishnu image was also found which has been placed in a tree shrine.

Local legends mention that most of the images and architectural fragments enshrined in the temples had been found in the *pokhar*, where they had been hidden during the Islamic raids. Once the images were recovered they were re-consecrated in the new shrines, sometimes with brand-new identities. In this process of relocation, often the images once recovered were placed under trees, for preservation and temporary worship; later shrines were often built at the same spot under the tree.

Legends of Rukmini once again reverberate at the tank to provide credibility and history to this sacred complex. As per living legends, Raja Sishupala, who wanted to marry Rukmini, contracted leprosy and came to Bargaon to the Suraj *pokhar*, when it was still a small patch of water. He accidently washed his hands in this water and his leprosy was cured. He then commissioned the digging of the *pokhar*. When the *pokhar* was dug up the icons now enshrined in the temples appeared. These scultpures were first kept on the banks of the tank and worshipped. When they started to disappear and some were reported stolen they came to be installed in the different temples. What appears interesting in the story is how a history of the site and the shrines has been recreated around the mythology behind the construction of the temples. It also highlights the fact that the Suraj *pokhar* is man-made and was dug out at a time in historical memory.

On the west bank of the tank are two *Shivalayas* with black basalt lingas, *nandi* (bull which represents Shiva's vehicle) and several architectural fragments under worship, including what would have been temple pillars, beams of gates and pedestals of images, all anointed.

To the east of the tank are a series of east-facing temples dedicated to Kali, Ram-Lakshmana-Janaki and Shankara-Parvati. Just before approaching this complex of shrines, in front of the Kali temple, a life-size standing Vishnu figure lies abandoned. It is badly damaged and only the upper body survives. The round back slab of the image contains a long inscription. Outside the Shankara-Parvati temple on a platform along the tank stray fragments of sculptures are found, including several lingas, Uma Mahesvara and what appears to be a votive temple, all duly anointed and worshipped. There is also a Mahisasuramardini temple further away from the tank with a black basalt image of the goddess under worship.

The bank of the Suraj *pokhar* is also the site of a travelling fair which goes around different villages in the sub-division and comes to Bargaon on Sundays. The biggest annual event here is however the bi-annual festival of Chhath celebrated all over Bihar in honour of the sun god. After making offerings to the setting and the rising sun, the fasting devotees walk up to the nearby Surya temple to worship his image in an anthropomorphic form.

Located in the centre of the village, the modern temple of the sun god contains a collection of old sculptures, Buddhist and Hindu, obviously taken from the ruins. The temple faces the north, unlike other Surya temples. The principal image is of Surya placed on a platform in the middle of the shrine and protected with iron grills. On the left of the Surya image is an Avalokitesvara figure worshipped as Chhath Maata. Below is an image of a female on a lion, worshipped as Singha Vahini, which is another name for Durga. Fixed on the wall behind are Vishnu, Satyanarayan and Bholeshankar, the last being a Buddha image in a *mundi* (miniature temple) niche.

Coming out of the sanctum, a *Shivalaya* on the left has a *panchamukhi* (Shiva linga with 5 human faces carved on it linga) as the principal image. On all the three walls a medley of large black basalt images are lined up along with several pillar fragments and votive stupas worshipped as lingas. The shrine is made of brick but stone pillars and thresholds from earlier temples have been reused. One of the most remarkable finds from Bargaon is a set of miniature temples with the image of Uma Mahesvara and a linga enshrined in the central niche. The original purpose of these temples, made of black basalt and granite, is difficult to explain.

Jagdishpur

The Rukministhan at Jagdishpur is another living instance of how myths have been reinvented where not only has the identity of the image been changed but a whole new mythological structure has been created around the shrine and the enshrined image. The site was visited by Buchanan, Cunningham and Broadley when the Buddha sculpture was still worshipped along with other Buddhist sculptures in an open-air tree shrine.[49] In the present day, this site is called Rukministhan and contains a single-cell shrine with a towering *shikhara*. This is a protected monument of the ASI. The shrine is associated with Vaishnava legends; the Buddha is worshipped as Krishna and the accompanying boddhisatvas as images of Rukmini. According to one local legend, the monastic complex, located about 2 kilometres away, was the Raja's palace where Rukmini lived. She came every day to worship at this shrine; as was pre-arranged Krishna abducted her from here. A second version of the legend tells that the monastic complex and the mound of the shrine are connected through a secret tunnel and after Krishna abducted Rukmini from her palace he used this tunnel to escape and reached the shrine. Religion and folklore have invested new meanings in landscape, identifying it with people and events from the past.

Whatever the mythological bearings, recent archaeological excavations at the site have revealed square-shaped cells, terracotta sealings and a number of votive stupas, besides other artefacts. The ASI believes it might be a sacred Buddhist structure or another monastery but not a part of the monastic complex:

> Based on preliminary assessment, it can be said it was a Buddhist religious structure, which might be a part of ancient Nalanda but not the university. The upper parts of the structure belong to the Pala period (8–12th century AD) and lower parts belong to the Gupta period (4–5th century AD). Nalanda ruins started getting developed around the BC sixth century and it took shape of the university around 4th century AD.[50]

Teliya Baba

Another interesting site of assimilation is the shrine of Teliya Baba at Bargaon. Located in between Temple 14 and Monastery 11 is a small brick enclosure which, though situated within the excavated site, does not belong to it at present. Inside is a colossal image of a seated Buddha, in *dharmachakrapravartan mudra* (also called "Turning the Wheel of Law" which symbolises the Buddha's first sermon at Sarnath), about 9 feet high, worshipped by the villagers as the Hindu god Bhairava or as Teliya Baba and by Thai pilgrims as Black Buddha. The Hindus believe the image to have unique healing power and they besmear it with oil as an act of worship. Other Buddhist images are also seen in the enclosure. It is not clear whether it marks the site of a temple which had disappeared long ago or was removed from one of the Buddhist temples nearby. This image must have been prominently situated and frequently attended to as an object of worship, for all the early explorers encountered it. Buchanan recorded it as "Baituk Bairobh", Cunningham referred to it as "Baithak Bhairav" and Broadley as "Telia Bhandar". Even to the present day the image is not a part of the protected enclosure of the ASI since the worshippers fought legally for full access to and control over the shrine. It can hence be argued that the Buddha has been assimilated within the Hindu traditions through the *puranic* pantheon as an *avatara* of Vishnu but is also closely associated with the Shaiva tradition as Bhairava.

Similar interchangeable identities are fairly common with several other images at Nalanda: for example, votive stupas in the villages are commonly worshipped as Shivalingas. Similarly, a stone image of Marici, the Buddhist goddess of dawn, close to Temple 14 is worshipped by the villagers as a Hindu deity.[51]

The area is also closely associated with Jain legends. The site of Kundalipur located about 1.5 kilometres away from the excavated area is believed to be the birthplace of Vardhamana Mahavir. During his wanderings, he spent the rainy months taking shelter near Nalanda. The spot, now marked by a modern shrine, is an important tourist destination and an annual fair is still organised here. A Jain temple dated to about 16th–17th century also survives at the Begumpur mound. Jain sculptures were found strewn on this mound both by Buchanan and Broadley. A significant number of Jain images have also been recovered from the excavated area.

Judging from the rich archaeological remains in the villages in the vicinity of the excavated area, one can imagine the enormity of the site and the vast amounts of sculptures and antiquities which have been displaced. During my field study, everywhere that I went I was told of the ancient sculptures from various shrines, particularly from tree shrines, which had begun to disappear before they were taken to the Patna Museum or the Nalanda Museum. Antiquities discovered during the course of colonial exploration and excavation at the sites are now stored at the Patna Museum, the National Museum and the Indian Museum, including parts of the Broadley collection, much of which came from Nalanda and its vicinity. Some bronzes from Nalanda were transferred to the Patna Museum in 1929 on a permanent loan while the Indian Museum houses antiquities which had been a part of Broadley's collection. The National Museum also acquired antiquities from Nalanda 'after selected works, including several Nalanda bronzes, were sent to

London for exhibition in 1947–48 honouring India's independence'.[52] When they were returned to India they came to form the core collection of the National Museum in New Delhi, which was established post-Independence in 1949, as representative of India's art and heritage. The images have travelled to be museum pieces serving as relics of the past and as objet d'art without any geographical or historical context and at the same time altering the understanding of the site itself.

Analysis and conclusions

Looking at the nature of remains in the villages around Nalanda, two issues become evident. First, is it judicious to understand Nalanda merely within the boundaries of the excavated area, as a Buddhist "university" detached from the villages in its vicinity? The entire locality forms an architectural complex of sacred structures of Buddhist, Hindu and Jain religions where structures and icons of one religion were readily appropriated and worshipped by another and where sacred space and areas of human inhabitation merged into each other. Second, my methodology suggests more meaningful ways of understanding how the monastic complex lived in popular memory after its supposed decline. Remains from the neighbouring villages pre-date the *Mahavihara* and continued to exist long after the monastery had collapsed. These villages are now dotted with several mounds which represent the remains of ancient structures.

By taking into account Hindu images in shrines and in museums and by corroborating sculptures with structures, I have attempted to trace this polytheism in religious practices in the villages around Nalanda. A more complex web of social networks and religious processes begin to take shape; religious images and iconography have been used to understand these religious processes. The sacred landscape is dotted with a series of shrines around the monastic complex in Nalanda, which have not been fully explored but form a part of a pilgrimage network of Buddhist, Jain and Hindu interests and provide connectivity and mobility both locally and within the region. It is important to locate the shrines and the other sacred enclosures within a social context to unravel the multiple levels at which sacred sites interacted with diverse communities. For instance the Rukministhan or the shrine of Teliya Baba are not frequented merely by Buddhists but by various communities of believers. When placed within a network of pilgrimage circuits between Gaya and Rajgir travelled not just by Buddhists but also by Jain and Hindu pilgrims, the much-popularised Buddhist character of the site seems only a part of the picture.

In a recent volume, Frederick Asher has suggested new possibilities of looking at Nalanda as stretching beyond the current limits of the monastic complex, which he defines as merely the area excavated by the ASI.[53] One of the methodologies he has relied on is the use of Landsat images, which have revealed water bodies, tanks or *pokharas* which surround Nalanda. The tanks fill up in the monsoons and suggest an alternative perimeter of the complex much beyond the excavated remains and also indicate a rich farming belt on which the monasteries would have sustained. Asher points out that the surrounding villages – Bargaon, Surajpur and Begumpur – would have all been part of the monastic complex as revealed by the rich corpus of antiquities still present in these villages. A second network Asher suggests is to place Nalanda, the monastery, within the lives of the greater population around it. He believes that 'In antiquity, as in the present, Nalanda did not exist as an isolated entity' but was dependent on a larger catchment area around it for people and for services.[54] In this light he views the built structures at Rajgir, Kundalipur and Pawapuri and traces patterns of artistic exchange and of a "secular urban" base. A third possibility which he suggests is to understand Nalanda within a network of other contemporary monasteries flourishing in the Magadha region at Uddandapura, Ghosrawan and Telhara, all located within a radius of 25 kilometres.[55]

Current historiography claims that the monastery at Nalanda received its deathblows at the hands of Muslim invaders somewhere around the 12th–13th centuries. Studying patterns of arrangement of shrines and icons in the present day indicate how the site survived in popular memory once the monastery had faded. One way of approaching the issue is through the *itihasa purana* tradition where new histories and mythologies of sites were represented and where older icons acquired new identities so that they could be integrated within the themes of the epics and the *puranas*.[56] The invisible strands of the myth created a sacred geography comprising of shrines, hills, tanks, trees and rocks through many centuries of redaction while living tradition such as *tirtha* (pilgrimage), festivals, fairs, poetry and literature ensured a continuous inhabitation of sacred sites. In the past, this tradition was ignored by historians because of its largely mythological content but a closer reading of myths gives a picture of the changing religious landscape.

In the villages around Nalanda, oral histories began to relate to the extensive ruins of the monastic complex, consisting mainly of earthen mounds strewn with bricks and carvings as representing the site of the ancient town of Kundalipur mentioned in the epics and the *puranas* as the capital of King Bhimaka of Vidarbha, the father of Rukmini, wife of Krishna.[57] The Rukministhana and Suraj *pokhar* became some of the sites for the re-enactment of these oral traditions. The Jain chronicles meanwhile associate Bargaon with the life of Vardhamana Mahavira. Hence though the *Mahavihara* ceased to exist, the memory of the site survived through various oral and ritual traditions.

Another powerful tendency seen in the period post 11th–12th centuries was the blurring of religious boundaries between Shaivism and Buddhism, resulting in the accommodation of religious and cultural elements of other religions including folk traditions and Tantricism. Buddhist monasticism began to decline such that in ritual and philosophy it came to grant to the laity a certain leniency of practice. The sacred landscape and its rituals were then reinvented and certain icons and shrines acquired identities which were neither Hindu nor Buddhist. This process of absorption is visible at multiple levels particularly in Bihar in rituals, architectural configuration and iconography. Tibetan sources refer to the five great *Mahaviharas* located at Vikramsila, Nalanda, Somapura or Paharpur (now in Bangladesh), Odantapura (the location of which is still undetermined), and Jagaddala, known for its *Vajrayana* preceptors.[58]

As discussed in an earlier section, the sculptural remains from the monastic complex at Nalanda – including many of those that are now in the Nalanda Museum – include a mix of Buddhist and Hindu images including Sarasvati, Surya, Vishnu, Lakshmi, Ganesha, Kubera, Kamadeva, Trilokavijaya, Heruka, Marichi, Tara, Vageshwari and so on. Excavations in what is believed to be the living quarters of the monks at Nalanda give a different kind of sculptural evidence: a number of small, bronze images were found, many of these still in wall niches, which indicates that they might have been meant for personal worship. Icons such as Trilokavijaya and Aparajita suggest the prevalence of Tantric rituals.

Accounts of the Chinese travellers such as I-Tsing have also attested to the presence of various shrines with colossal images of the Buddha and have elaborated upon the ritual worship of these images by priests.[59] The images from the monastic complex as well as those from the villages surrounding it are unusually large in size. The most apparent example of these is the Buddha image in the Rukministhan at Jagdishpur which I have discussed in the last section and which is more than 12 feet high. Some of the Hindu images available from Bargaon – such as a Parvati image in the Surya *mandir* and others now enshrined in temples around the Suraj *pokhar* – are all more than 5 feet in height (Figure 6.4). It has been suggested that such large images were meant

Figure 6.4 Parvati image and other sculptural fragments at Suraj *mandir*, Bargaon.

Source: Courtesy of the author

to be enshrined as the principal image of the sanctum.[60] Excavations at the temples in the monastic complex have unearthed partial remains of several such large stucco images such as the Buddha in *bhumispar-samudra* in Temple 12 and other broken stucco Buddhas in Temple 13 and Temple 14. The existence of these large images possibly also indicates the configuration of the site as per the Tantric mandala, with the colossal Buddhas being consecrated and worshipped in the cardinal directions as the guardians of the four quarters. Similar evidence emerges from the site of Tetrawan, located about 20 kilometres away from Nalanda, where another colossal Buddha in *bhumisparsamudra* figure, 9.5 feet high, is worshipped by locals of all faiths as Bullum Bhairava. Thus the immediate network of Nalanda is evident, but what is also obvious is the wider significance of the site.

An analysis of the symbiotic relationship between the monastery, the shrines and the villages – reinforced through mythologies, ritual praxis and religious motifs – invests new meanings into a "dead" World Heritage Site and situate it within present-day tourist and sacred networks. By attaching the Buddhist tag to Nalanda, the World Heritage listing has defined the limits and carved out the imaginary boundaries within which the site is located, much in the colonial fashion. As isolated monuments, devoid of human networks and participation, the Western conception of heritage as a homogenous category has been reimposed:

> Because values are in our minds and not inherent to objects, site valuation is fundamentally an extrinsic process. . . . At a specific moment in time, varying individuals or communities can also associate different values with a specific site that as a result concentrates webs of meaning.[61]

There is hence no "universal meaning" of a monument: it is rather open to interpretations, representations and forms of preservations. Values are created, based on a certain phase of knowledge production of a specific site as had been the case in colonial India. At the same time, identifying its "Outstanding Universal Value" to humanity changes the local meaning and significance of the site such that the excavated complex of Nalanda "the monastery" now stands detached from its immediate hinterland. It is hence important to trace the process of continuity and change, to show that sites do not die but get reinvented through ritual and faith so that we can breathe life into reliquaries of the past and not perceive them merely as heritage structures.

Notes

1 'Archaeological Site of Nalanda Mahavihara (Nalanda University) at Nalanda, Bihar', http://whc.unesco.org/en/list/1502 (accessed on 25 May 2017).
2 V. H. Jackson (ed.), *Journal of Francis Buchanan Kept During the Survey of the District of Patna and Gaya in 1811–12*, Bihar, Orissa and Patna: Superintendent Government Printing, 1926, p. 86.
3 Jackson (ed.), *Journal of Francis Buchanan*, p. 76.
4 Jackson (ed.), *Journal of Francis Buchanan*, p. 79.
5 Jackson (ed.), *Journal of Francis Buchanan*, pp. 100–101.
6 Montgomery Martin, *The History, Antiquities, Topography and Statistics of Eastern India*, Vol. I, London, Behar and Shahabad: W. H. Allen & Co., 1838, p. 95.
7 Archaeological Survey of India, *Report for the Years 1861–65*, Shimla: Government Central Press, 1871, p. 28.
8 Metal image inscription of Gopaladeva, year one, in Archaeological Survey of India, *Report for the Years 1861–65*, p. 36.
9 Metal image inscription of Devapaladeva, year three, in Hiranand Shastri (ed.), *Nalanda and Its Epigraphic Material, Memoirs of the Archaeological Survey of India, No. 66*, Calcutta: Government of India Press, 1942, p. 87.

10 Metal image inscription of Devapaladeva, year three, in Hiranand Shastri (ed.), *Nalanda and Its Epigraphic Material*, p. 29.

11 Even in the present day the village of Jagdishpur is also known as Rukministhan.

12 Archaeological Survey of India, *Report for the Years 1861–65*, p. 28.

13 Cunningham recorded the exact measurements of the sculpture as 15 feet tall and 9.5 feet wide, Archaeological Survey of India, *Report for the Years 1861–65*, p. 28.

14 Archaeological Survey of India, *Report for the Years 1861–65*, p. 30.

15 Archaeological Survey of India, *Report for the Years 1861–65*, p. 34.

16 Archaeological Survey of India, *Report for the Years 1861–65*, p. 36.

17 A. M. Broadley, 'The Buddhistic Remains of Bihar', *Journal of Asiatic Society of Bengal*, 1872, 41(3): 209–312.

18 Broadley, 'The Buddhistic Remains of Bihar', p. 239.

19 Broadley, 'The Buddhistic Remains of Bihar', pp. 209–312.

20 Broadley published a separate pamphlet on his studies at Bargaon, Nalanda: 'Ruins of the Nalanda Monasteries at Bargaon, Sub-division Bihar', Zillah Patna, Calcutta, 1872.

21 A. M. Broadley, *The Buddhistic Remains of Bihar*, first published in 1872, this edition, Varanasi: Bharati Prakashan, 1979, p. 95.

22 Ibid.

23 Ibid.

24 Broadley, *The Buddhistic Remains of Bihar*, p. 96.

25 Ibid.

26 Ibid.

27 Education Department File number XI E-47 of 1922, Bihar State Archives, Patna.

28 As noted in the display labels at the museum.

29 Frederick M. Asher, *Nalanda: Situating the Great Monastery*, Mumbai: Marg, 2015, p. 96.

30 A. Ghosh, *A Guide to Nalanda*, New Delhi: Manager of Publication, first published in 1939, this edition 1950.

31 By far the richest collection is that of stone and bronze images of gods and goddesses of the Buddhist, and, in a few cases, of the Brahmanical pantheon. The images are found in an abundance in the monasteries where they were worshipped and in all probabilities manufactured.

> (Ghosh, A Guide to Nalanda, p. 21)

32 The presence of not a negligible number of Brahmanical images in the centre of Buddhist theology and ritual is intriguing. Probably their introduction and existence were tolerated, but it must be remembered that this was the age when the Buddhist were conceiving and erecting such deities a Trilokyavijaya trampling on Shiva and Parvati. It is no doubt that there were mutual exchange and borrowing of deities, but it is not possible to think that a Brahmanical deity whose image we find at Nalanda was ever absorbed in the Buddhist pantheon

> (Ghosh, A Guide to Nalanda, p. 21)

33 Broadley, *The Buddhistic Remains of Bihar*, p. 94.

34 Archaeological Survey of India, *Report for the Years 1861–65*, p. 28.

35 J. Kamini, Malatesh Kulkarni, V. Raghavaswamy, P. S. Roy and P. K. Mishra, 'CARTOSAT-1 Views the Nalanda Buddhist Ruins', *Current Science*, July 2007, 93(2): 136.

36 M. B. Rajani, 'The Expanse of Archaeological Remains at Nalanda: A Study Using Remote Sensing and GIS', *Archives of Asian Art*, 2016, 66(1): 1–23.

37 Jackson (ed.), *Journal of Francis Buchanan*, p. 96 and Broadley, *The Buddhistic Remains of Bihar*, p. 94.

38 Archaeological Survey of India, *Report for the Years 1861–65*, p. 34.

39 Archaeological Survey of India, *Report for the Years 1861–65*, p. 36.

40 Archaeological Survey of India, *Report for the Years 1861–65*, p. 30.

41 B. N. Misra, *Nalanda*, Vol. 1–3, New Delhi: BR Publishing Corporation, 2008, Vol. 1, p. 170.

42 D. R. Patil, *Antiquarian Remains in Bihar*, Patna: K. P. Jayaswal Research Institute, 1963 (p. 307):

> There is strong reason to believe that many of the ponds or pokhras were not natural sheets of water nor were they artificially excavated to serve as ponds and reservoirs, but were excavated mainly for the earth required to manufacture the enormous quantity of bricks needed to raise the huge buildings of temples and monasteries at the site.

43 Rajani, 'The Expanse of Archaeological Remains at Nalanda', pp. 1–23.

44 Patil, *Antiquarian Remains in Bihar*, p. 328.

45 Misra, *Nalanda*, p. 264 and Dhaky Meister and Krishna Deva (eds), *Encyclopedia of Temple Architecture*, p. 110.

46 AM Broadley, *The Buddhistic Remains of Bihar,* p. 95.

47 Jackson (ed.), *Journal of Francis Buchanan*, p. 100.

48 There have been several independent studies on Nalanda such as of H. D. Sankalia (1934), A. Ghosh (1939), Mary Stewart (1989), B. N. Mishra (2008), C. Mani (2008) etc.

49 My visit to the site in April 2011 confirmed the same. The image is still in active worship evident from the anointments and offerings of hibiscus flowers to the image. A number of smaller Buddhist figures are placed next to the main image. The priest of the shrine told me that the present temple had been built in 1973 and that there were several other images at the site but once they started getting stolen the remaining were removed to the Patna Museum and the Nalanda Museum. There is also a well located nearby, on top of the mound.

50 'Buddhist Mound found near Nalanda ruins: ASI rules out any direct link with artefacts excavated close to ancient University', *The Telegraph*, 15 June 2015, www.telegraphindia.com/1150615/jsp/bihar/story_25729.jsp (accessed on 30 May 2017).

51 Ghosh, *A Guide to Nalanda*, p. 20.

52 Asher, *Nalanda*, p. 96.

53 Asher, *Nalanda*.

54 Ibid, p. 104.

55 Asher, *Nalanda*, p. 110.

56 The core of the *itihasa purana* tradition, which comprises the epics and the *puranas,* was initially preserved through memory and oral renditions and can be dated to the seventh century BCE. The *itihasa purana* tradition

154 *Salila Kulshreshtha*

has three main constituents: myth, genealogy and historical narrative. The remote past is described through myths; the more immediate past is recorded in the form of genealogies, while historical literature provides historical narratives, Ray, *Return of the Buddha*, p. 23.

57 Patil, *Antiquarian Remains in Bihar*, p. 301.
58 Ray, *Return of the Buddha*, p. 187.
59 H. D. Sankalia, *The University of Nalanda*, New Delhi: Oriental Publishers, 1972, p. 128.
60 Claudine Bautze-Picron, 'Crying Leaves: "Some Remarks on the Art of Pala India (8th–12th Centuries) and Its International Legacy"', *East and West*, December 1993, 43(1/4): 277–294.
61 Sophia Labadi, *UNESCO, Cultural Heritage and Outstanding Universal Value*, Walnut Creek, CA: AltaMira Press, 2013, p. 15.

Bibliography

Asher, Frederick M., *Nalanda: Situating the Great Monastery*, Mumbai: Marg, 2015.

Bautze-Picron, C., 'Buddhist Slabs From Nalanda', in *Berliner Indologische Studien*, Band 6, Reinbek, 1991, pp. 81–100.

Broadley, A. M., 'The Buddhistic Remains of Bihar,' *Journal of Asiatic Society of Bengal*, 1872, 41(3).

Broadley, A. M., *The Buddhistic Remains of Bihar*, first published in 1872, this edition, Varanasi: Bharati Prakashan, 1979.

Cunningham, Alexander, *Archaeological Survey of India, Report for the Years 1861–65*, Shimla: Government Central Press, 1871.

Ghosh, A., *A Guide to Nalanda*, 3rd edition, Calcutta: Manager of Publication Delhi, Government of India Press, 1950.

Huntington, Susan, *The Pala-Sena Schools of Sculpture*, Leiden: EJ Brill, 1984.

Jackson, V. H. (ed.), *Journal of Francis Buchanan Kept During the Survey of the District of Patna and Gaya in 1811–12*, Bihar, Orissa and Patna: Superintendent Government Printing, 1926.

Kamini, J., Malatesh Kulkarni, V. Raghavswamy, P. S. Roy, and P. K. Mishra, 'CARTOSAT-1 Views the Nalanda Buddhist Ruins', *Current Science*, July 2001, 93(2): 136.

Kumar, Brajmohan, *Archaeology of Pataliputra and Nalanda*, New Delhi: Ramanand Vidya Bhavan, 1987.

Labadi, Sophia, *UNESCO, Cultural Heritage, and Outstanding Universal Value: Value Based Analysis of the World Heritage and Intangible Cultural Heritage Conventions*, Lanham, MD: AltaMira Press, 2012.

Leoshko, Janice, *Sacred Traces: British Explorations of Buddhism in South Asia*, Fanham: Ashgate Publishers, 2003.

Mani, C. (ed.), *The Heritage of Nalanda*, New Delhi: Aryan Books International and Asoka Mission, 2008.

Misra, B. N., *Nalanda*, 3 Vols., New Delhi: BR Publishing Corporation, 2008.

Patil, D. R., *The Antiquarian Remains in Bihar*, Patna: KP Jayaswal Research Institute, 1963.

Rajani, M. B., 'The Expanse of Archaeological Remains at Nalanda: A Study Using Remote Sensing and GIS', *Archives of Asian Art*, 2016, 66(1): 1–23.

Ray, H. P. (ed.), *Sacred Landscapes in Asia: Shared Traditions, Multiple Identities*, New Delhi: Manohar, 2007.

Ray, H. P., 'From Multi-Religious Sites to Mono-Religious Monuments in South Asia: The Colonial Legacy of Heritage Management', in Patrick Daley and Tim Winter (eds), *The Routledge Handbook of Heritage in Asia*, London: Routledge, 2012.

Ray, H. P., *The Return of the Buddha: Ancient Symbols for a New Nation*, London: Routledge, 2014.

Ray, H. P., and Carla M. Sinopoli (eds), *Archaeology as History in Early South Asia*, New Delhi: ICHR and Aryan Books International, 2004.

Sankalia, H. D., *The University of Nalanda*, New Delhi: Oriental Publishers, 1972.

Shastri, Hiranand (ed.), *Nalanda and It's Epigraphic Material, Memoirs of the Archaeological Survey of India*, No. 66, Calcutta: Government of India Press, 1942.

Singh, Upinder, *The Discovery of Ancient India: Early Archaeologists and the Beginnings of Archaeology*, New Delhi: Permanent Black, 2004.

Sinha, B. P., 'Archaeology in Bihar', KP Jayaswal Memorial Lecture Series, Vol. V, KP Jayaswal Research Institute, Patna, 1988.

Stewart, Mary L., 'Nalanda Mahavihara: A Study of an Indian Pala Period Buddhist Site and British Historical Archaeology, 1861–1938', British Archaeological Reports International Series 529, Oxford, 1989.

7 The Qutub Minar complex and the village of Mehrauli

Multiple meanings in monuments

Swapna Liddle

The mediaeval history of Mehrauli

The Qutub Minar is an iconic landmark of Delhi, the distinctive silhouette of this 72.5-metre-high stone tower immediately identifiable with the city. This more than 800-year-old structure and the complex of buildings within which it stands, are preserved as a national monument and UNESCO World Heritage site. For the hundreds of visitors that come here daily, here lies the story of Delhi, and of North India, in the period following the "Muslim conquest".

The Qutub Minar lies in Mehrauli, the site of Delhi's oldest fortified city, built by the Tomar Rajput ruler Anangpal in the middle of the 11th century CE. A century later, the fort was enlarged by the Chauhans, who replaced the Tomars as the rulers of Delhi. This city was conquered in 1192 by the forces of Muizzuddin Ghuri, the ruler of Ghazni in Afghanistan. After the conquest, Muizzuddin left Delhi in the care of his military slave general Qutubuddin Aibak, an ethnic Turk.[1]

One of the earliest acts of this new dispensation was to build a congregational mosque for the use of the many Muslims, mainly soldiers, who now came to be stationed in Delhi. An inscription on the inner lintel of the eastern gateway reads, 'The materials of 27 temples, on each of which 2,000,000 Dehliwals [the currency of the time] had been spent, were used in [the construction of] this mosque'.[2] These temples were clearly destroyed by the conquerors, in the moment of conquest. It is likely that these places of worship were selectively chosen, being temples that were associated with the previous regime, such as the grand Jain temple built by the Jain merchant Sahu Nattal in the year 1132.[3] Other shrines that enjoyed purely popular support, such as the Dadabari Jain temple and the Yogmaya temple, were spared.[4]

They are still to be found in Mehrauli, and are popular, living places of worship.

In practical terms, the "destruction" of temples was more in the nature of a dismantling, as pillar columns and capitals were carefully taken apart, only to be put together in the construction of the mosque. Some of the features of the human and divine figures carved into the temple stones were defaced, to make them conform to the Islamic proscription of idolatry. These pillars and other stone members such as beams and lintels can still be seen in the mosque. The dismantling of the temples and the putting together of the mosque was no doubt the work of Indian builders – workers and their supervisors. These were the same skilled craftspeople who had executed the original temples through several generations. Moreover, these builders exercised a good deal of their own discretion when it came to deciding how the mosque would be made. As a result, they often used conventions they were familiar with. For instance, unlike in most mosques, where the western side alone is emphasised, in the Qutub mosque the eastern side was provided with four rows of pillars, making it more prominent, as would have been the case in a temple.[5]

Not all the stones used in the mosque were recycled elements from older structures. Some were carved afresh, such as the arches and lintels over the doorways. In this case too, the stone carvers exercised their independent discretion to a significant extent. They placed a *kirti mukha* over the lintel of the eastern doorway. This symbol, an important motif in Hindu temples, represents the legendary self-devouring demon created by Shiva, and was popularly used as a device to ward off the evil eye. It was not a symbol that carried any significance for the Turk military commander who commissioned the mosque. It was the work of the carvers alone (Figure 7.1).

A similar approach can be seen in the ornamentation of the towering screen of arches which indicated the *qibla*, the direction of Mecca, and was built some years after the conquest. Here too the builders carved several traditional Indian symbols. One was the *purna kalash* or *purna ghata*, a vessel symbolising abundance and the source of life, and an important ritual element in Hindu worship. The carvers were familiar with the practice of carving idols of Ganga and Yamuna, the river goddesses, on columns flanking the entrance ways to temples. Since their new clients prohibited the carving of idols, they instead used symbolic representations of the goddesses, such as the tail of the *makar*, the mythical crocodile that is believed to be the vehicle of Ganga (Figures 7.2 and 7.3).

Figure 7.1 Detail of the eastern doorway of the Qutub mosque.

Source: Courtesy of the author

In the making of the Qutub mosque, the Indian construction workers and their supervisors were also experimenting with some fresh
motifs and techniques that their new clients were introducing them
to. The carvers were asked to incorporate inscriptions in the Arabic
script into the surface decoration of the structure. This too they interpreted in their own way. In the panels of calligraphy on the screen,
they carved lotus vines. The lotus, again, was a symbol that had a particular significance in the Indian tradition. In the Qutub Minar itself,
begun during the reign of Qutubuddin Aibak, a similar idiom was
employed.

An important new idea that the Turks brought to India was that of
the arch and the dome. Technologically this was a radical departure
from the Indian tradition of covering space using trabeate construction, based on columns and beams. The earliest experiments with arch
making in Delhi resulted in a corbelled arch, which had the shape of
a pointed true arch but not its structural strength. It was based on
techniques the builders were already familiar with. Similarly they built
cone-shaped domes, using principles that would have been applied to

Figure 7.2 Ganga on the *makar*, Sas Bahu temple, Gwalior.

Source: Courtesy of the author

Figure 7.3 Symbolic representation of Ganga on the *qibla* arches of the Qutub
  mosque.

Source: Courtesy of the author

make *shikharas*, or spires, of temples, instead of arch-making principles, which would have resulted in rounded domes.

The reason for building a corbelled "false arch" was simple: the Indian builders were not familiar with the structure of a true arch, and there were probably no builders among the Turk conquerors to guide them. What is more difficult to understand is why many decades would elapse before the true arch began to be used in the architecture of Mehrauli. Emperor Iltutmish built an extension of the mosque in the 1230s, more than 30 years after the original structure was complete, and he built his own tomb next to it. In neither of them can the true arch be seen. By this time, more than a generation after the conquest, no doubt people other than soldiers were pouring into Delhi from Central and West Asia. In fact, we can see the work of sophisticated calligraphers who designed some of the surface decoration of the mosque extension – for instance in the inscriptions which use sophisticated styles such as the knotted Kufic script. It is also important to point out that decoration of this structure with so-called hallmarks of "Islamic" ornament – calligraphy, floral and geometric motifs, and arabesques – did not preclude the continued use of the *kalash* and lotus as major elements in the design of the *qibla* arches (Figure 7.4).

Why did builders coming into Delhi from Central and West Asia not correct the arch the Indian builders were producing? This tells us something important about the position of the Indian masons under the sultanate. They apparently had enough control over the decision-making process that while they were willing to let some of the surface decoration be designed by the newcomers, they kept a tight hold on their basic building processes and would not be dictated to. It took some three-quarters of a century before the Indian builders were convinced of the soundness of the arch, and then adopted it. In Delhi the earliest true arch is in evidence in the tomb of Balban, close to the Qutub Complex, which was built in the 1280s.

The point that is being made is that Indian builders were not mere unwilling instruments of the new regime. The histories of conquest are often written in the rhetoric of war, bloodshed and destruction. No doubt these were an inevitable outcome of the replacement of one regime by another, but we must not assume that it was a story of endless conflict at every level. It is easy to overstate the case when the new rulers were culturally different from the old. As the case of the selective temple destruction described suggests, the conquerors probably made clear distinctions between the *regime* they had defeated and displaced, and the *people* whom they hoped to rule. On their part, the

Figure 7.4 *Kalash* on Iltutmish's extension of the Qutub mosque.

Source: Courtesy of the author

latter were interested in working out their own relationships with the newcomers.

Professionals, like the builders of the Qutub mosque, were willing to lend their services to the new patrons, just as they had lent them to earlier state patrons. Therefore, even in the early stages of the interaction, we see a significant degree of co-operation and compromise in the creation of these architectural masterpieces.[6] The Indian workpeople moreover retained their faith as well as their professional practices. This is attested to, for instance, by the invocation of Vishwakarma, the ultimate source of architectural knowledge, in a Sanskrit inscription on the Qutub Minar, dating from 1369, on a stone used in a repair project of that date. They also continued to play a leading role in future projects within the complex. The same inscription mentions that the architect was 'the maternal grandson of the son of Chahadadevapala'.[7]

With the passage of time, these early experiments in structure and ornamentation would lead to a more thorough cultural assimilation. Mention has been made of the use of traditional symbols such as the *purna kalash* and the lotus used by the stone carvers in profusion in the new stones they were carving for the Qutub mosque. It is of course understandable that in the first few years of the sultanate, the Turk patrons understood little of the symbolism of these motifs. However, certainly within a generation or so they must have realised the sacred significance of these symbols in the traditional culture, and they were still happy to have them incorporated into their own religious architecture, such as mosques and tombs. These symbols continued to be used till late Mughal times.

Over the centuries many more motifs were added to the repertoire of surface decoration on buildings. These included *torana* arches, *padmalata* bands, and *padmajala* panels. The finials of mosques and tombs began to be modelled on the *amalaka* and *kalash* finials on temple shikharas. An *amalaka* can be seen on the early 14th-century Alai Darwaza, built by Alauddin Khilji as a monumental entrance to the Qutub mosque. The *kalash* forms part of the finial of the nearby tomb of Imam Zamin, dating from the 1530s. Another motif that was given a prominent place was the six-pointed star. Though stars of different configurations were used in decorative schemes in the West and Central Asian tradition, the six-pointed star had a spiritual significance in the Indian tradition in that it represented the union of Shiva and Shakti. The six-pointed star was given pride of place in important monuments such as Alai Darwaza in the Qutub Complex, and elsewhere, such as in Humayun's Tomb (Figure 7.5).

How do we account for the Muslim patrons' acceptance of Indic sacred and other popular motifs into the artistic idiom of the buildings

Figure 7.5 Kalash and *amalaka* finial of the tomb of Ghiyasuddin Tughlaq in
Tughlaqabad, Delhi.

Source: Courtesy of the author

they commissioned? Part of the answer must lie in the strategy and the
attitude towards the primarily non-Muslim population of India that
had been adopted in the early years of the sultanate. It has been noted
that the rulers made it clear that their quarrel was only with those
who questioned their political power and their right to collect revenue. In the interests of peace, it served them well to adopt a tolerant
policy towards the social and religious practices of their non-Muslim
subjects, and in fact to even accommodate and adopt some of them.
In this, they were helped by the preaching and practice of many of the
Sufi saints of the time.[8] This is best illustrated by a remarkable *mihrab*
(the arch indicating *qibla*) in the tomb of the saint Maulana Jamali,
who died in 1535 and is buried in Mehrauli. In its centre is the carving
of a *kalash* on which is written "Allah" (Figure 7.6).

Figure 7.6 *Qibla* of the tomb of Maulana Jamali.

Source: Courtesy of the author

State control and reading of a monument

Mehrauli did not remain the capital of the sultanate for long. The centre of power moved away, to other sites in Delhi, and also at times away from Delhi altogether. Yet Mehrauli continued to be inhabited, and it flourished. A significant explanation for its continued importance lay in its shrines: the Yogmaya temple, the Dadabari Jain temple, and the shrine or *dargah* of Qutubuddin Bakhtiyar Kaki, the Sufi saint who died and was buried there in 1235. These drew visitors from outside, including rich patrons, often emperors like Sher Shah Suri in the 16th century and Bahadur Shah I in the 18th century, who built mosques, gates and other structures here. Repairs to the older buildings, such as the Qutub Minar, which was prone to damage from earthquakes and lightning strikes, were also regularly carried out.

By the 19th century, several people from the city of Delhi (the Mughal city of Shahjahanabad, now popularly known as Old Delhi), had built second homes here and laid out gardens. Part of the reason for Mehrauli's popularity was its reputation for a more healthful climate and location than that of Shahjahanabad. People from the city would therefore come here regularly for a "change of air". There were instances of many people leaving for Mehrauli when there was a contagious illness in the city.[9] In fact, soon some people were beginning to complain that precisely because Mehrauli had become so popular it was visited by many people and, as a result, was becoming littered and polluted.[10]

The Mughal royal family also liked to make prolonged visits to Mehrauli, that usually lasted a few days. Emperor Akbar II built a palace in Mehrauli, next to the *dargah* of Qutubuddin Bakhtiyar Kaki. This palace later came to be known as Zafar Mahal, after his son Bahadur Shah Zafar extended it in the 1840s. Both emperors also added pavilions next to a man-made waterfall (known as the Jharna) that came from the Shamsi reservoir, which dated from the 13th century.[11] On visits to Mehrauli, the emperor was usually accompanied by many members of his extended family and members of his court. Areas were screened from the public eye by temporary constructions so the ladies of the family could enjoy the open space. Parties were organised in the mango grove nearby, in which swings were strung up from the trees. Open air entertainments of singing and dance were held for the royal visitors at the Jharna, which was also the place where the begums bathed.[12] The Mughal emperors were also patrons of a local festival, the Phoolwalon ki Sair, or festival of the flower-sellers. This annual festival, which began in the early 19th century, involved simultaneous veneration of and offerings at the *dargah* of Qutubuddin Bakhtiyar Kaki and the temple of Yogmaya.

By the early 19th century, Delhi was under the control of the British East India Company, while the Mughal emperor was soon reduced to less than a figurehead. British officials soon began to take an interest in Mehrauli, and this interest was at two levels. The first step taken was the repair of the Qutub Minar, which had been damaged in an earthquake in 1803. A project for repair and renovation was undertaken in the late 1820s by army engineers led by Major Robert Smith. The work, however, revealed a lack of understanding and sensitivity to a remarkable degree. The facing stones that had fallen down were put back, but randomly, so that the inscriptions were now jumbled and unreadable. Numerous additions were made that did not match the original structure in either style or spirit: balcony balustrades on four stories, an iron and brass balustrade for the fifth storey, a stone pavilion, and over it all, an octagonal wooden cupola supporting a flagstaff.[13] Fanny Parks, an Englishwoman visiting the monument in 1938, criticised the new additions. The pavilion on top was disproportionately heavy, she said. Moreover, 'Not content with this, he [Smith] placed an umbrella of Chinese form on the top of the pavilion; it was not destined to remain – the lightning struck it off, as if indignant at the profanation'.[14]

Neither was the work at the Qutub Minar only about repairs and additions to the structures. The area around the *minar* and mosque was at the time covered with mud huts and the remains of several other structures dating from various periods in the history of the complex. These were completely demolished, leaving only the more attractive structures. The resulting debris was piled up into 'ornamental mounds and seats of view', on at least one of which a pavilion was constructed. The resulting landscaped open ground was planted with trees and shrubs, and roads were laid to provide good access to the area from its immediate neighbourhood.

The buildings that had been cleared away from the area were not all old and abandoned structures. Some of them clearly were the houses of religious mendicants and *khadims* (keepers of the *dargah*) and steps were taken to ensure that they would be unable to return to the area. The area was to be guarded and maintained by a permanent staff and elaborate rules were promulgated regarding access to the buildings and the space around them. There were clear injunctions against defacing the buildings and plantations, and against occupying any part of the grounds or buildings for any length of time. Camping was to be allowed only on areas specifically demarcated for the purpose. Finally, notices in Persian and English were laid down:

> Groups of natives of the lower classes are not to ascend the minar in a number exceeding ten at a time, and are not allowed to make any stay in the minar. No fakirs, beggars or vagrants are allowed to take up their post or demand money about the Kootub.[15]

In the coming years a wider area around the Qutub would be taken over and modified by individual British officials for their private use. A certain civil servant by the name of Blake took over the tomb of Adham Khan, the 16th-century Mughal noble, and converted it into a home. Some years later, in 1844, the highest British official in Delhi at the time, Thomas Metcalfe, converted the nearby tomb of Adham Khan's brother, Quli Khan, into a second home, calling it "Dilku-sha". Metcalfe seems to have bought the tomb from a moneylender to whom it had been mortgaged.[16] Rooms were added around the central chamber and other buildings in the vicinity were modified to provide accommodation for staff and for other ancillary facilities. The visual landscape was changed by the construction of pavillions and follies in the surrounding open space – these included two ziggurat/pyramid-like structures and a couple of canopies in imitation of old *chhatris*.

While British officials spent time in Mehrauli, they did not connect socially to the life of the place in the way in which the Mughal royal family did. The latter supported the shrine and sponsored events like the annual Phulwalon ki Sair. Individual British officers also visited the fair, but purely as spectators[17] – and as Syed Ahmad Khan mentioned regretfully, there was no holiday in government offices on the occasion, to enable government employees to visit Qutub on the occasion![18]

The 19th century did not merely see growing control of the monuments of Mehrauli by the British colonial state and its officials. From the mid-19th century onwards, scholarly attention was being turned to the study and interpretation of these monuments. The earliest initiatives were undertaken under the aegis of the Delhi Archaeological Society, which was composed of members of Delhi's official and intellectual elite. A paper presented by Syed Ahmad Khan to the society, and the subsequent entry in his 1847 publicaton, *Asarussanadid*, was the first among a series of archaeological reports that would shape the modern understanding of the site.

Syed Ahmad Khan's work is the first reference we have of the name "Quwwat-ul-Islam" being applied to the Qutub Mosque: in older texts as well as in the mosque's inscriptions it is referred to simply as the

"Jami Masjid" (the congregational mosque). *Quwwat-ul-islam*, literally "might of Islam", is a term that appears to have been a corruption of *qubbat-ul-Islam*, or "sanctuary of Islam", a term used in the 13th century for the city of Delhi itself. The name Quwwat-ul-Islam, when applied to the Qutub mosque, lays a degree of emphasis on the iconoclastic origins of the mosque that was probably not intended by the original builders and worshippers. Some of Syed Ahmad's other assertions, such as that the Qutub Minar, or at least its first storey, were constructed by Prithviraj Chauhan, who ruled Delhi till overthrown by Mohammad Ghuri, have not been generally accepted, with exceptions that will be discussed later. However, his use of the name Quwwat-ul-Islam became an unquestioned fact repeated by most official documents of the Archaeological Survey of India.[19]

These later reports of the Archaeological Survey of India, from that of Alexander Cunningham in 1862–1865 to J. A. Page in 1926, would take the theme of iconoclasm further, positing barbarism and rapacity as the defining features of mediaeval Muslims in general. The classification of different decorative motifs and forms within the complex into the rigidly defined categories of "Hindu" and "Muslim" architecture was another assumption perpetuated in these studies. In this reading, the carving of a *kalash* decorated with Quranic verses, in Iltutmish's extension of the Qutub mosque, is an anachronism, and glossed over. This official view, enshrined in the reports, went on to inform the most widely distributed guidebooks on the site. One of these guides, written by Page himself, further went on to paint a picture of the Hindu artisans working on the mosque in a subservient role, entirely directed by their Muslim masters, a scenario that has been challenged in this chapter.[20]

Mehrauli and its monuments in the popular perception

To go back to the beginning of the 19th century, how did those living in Delhi, and in Mehrauli in particular, react to the appropriation of these mosques and tombs of the locality by the colonial state, whether as exclusive private property, or as a state controlled "monument"? No clear evidence remains of "native" reactions to the measures undertaken to monumentalise the Qutub site. We may imagine that the residents of Mehrauli might have felt some resentment at being excluded from the immediate vicinity of the Qutub, and the turning of a site that was earlier integrated with the locality of Mehrauli into a space set apart and controlled by alien rulers. Though there are no explicit statements to be found in this regard, we do find evidence of some opposition of a covert sort. This was expressed soon after the repairs of the 1820s were completed, with the new balustrade being damaged by an anonymous vandal in, as the official report put it, "mere wantonness".[21] In the case of Metcalfe's appropriation of Quli Khan's tomb as a private residence, our only clue to the popular sentiments about such a modification and use of a tomb is an indirect one, and from Syed Ahmad Khan himself. When Syed Ahmad Khan wrote *Asarussana-did* shortly after the house was completed, he praised it highly, even applying to it the hyperbolic verse from Amir Khusro – 'If there be a paradise upon earth, it is this, it is this, it is this'. Yet he diplomatically left out all reference to the fact that it had originally been a tomb. He evidently felt that this was a sensitive subject, on which he, as a government servant, could not comment negatively, so he chose to keep quiet.[22]

Equally difficult to recover is the impact that the official reading of the Qutub complex, and more specifically the Quwwat-ul-Islam mosque, has had on popular opinion of this monument. The signage at the site, as well as the official guidebooks available today, do not emphasise the theme of Islamic conquest. They do however continue the use of the name "Quwwat-ul-Islam", which implicitly underlines the mosque as a symbol of the "Muslim" conquest of India. The descriptions of the architecture of the various structures also uncritically follow the colonial understanding of a dichotomy between Hindu and Muslim forms and styles. Most of the standard resources available even to students of architecture do not depart from this narrative.

The absence of an alternate reading of the structures within the Qutub complex, leaves the site open to interpretation in terms that elaborate on the assumptions implicit in the name of the mosque, and in the categorisation of motifs as Hindu and Muslim. Guides leading tourists around the complex point out the iconoclasm supposedly self-evident in the pillars used in the mosque's construction. It has also been argued, by those with a Hindutva agenda, that the Qutub Minar itself is a "Hindu" monument, "defaced" and converted into a "Muslim" one. Such arguments, though not endorsed by the ASI or any reputed scholars, have led on occasion to calls for reclamation of the site for Hindu worship. Such a protest took place in 2000, leading to the arrest of several protestors who were demanding that they be let into the complex to "liberate" the idols "trapped" within the mosque.[23]

The narrative of Muslim desecration and the resulting calls for retribution have left their impact in Mehrauli village too, which lies adjacent to the Qutub complex, of which the structures of the monument were once an integral part. In September 1947, in the shadow of Partition, the shrine of Qutubuddin Bakhtiyar Kaki was vandalised and all the Muslims of the locality were driven out. The status quo was eventually restored after some months, when Gandhiji specifically mentioned the restoration of the shrine to Muslims as one of the conditions for breaking his fast of January 1948, his last.[24] However, a 16th-century gateway to the shrine has since been appropriated by a Sikh *gurudwara* and rededicated as a shrine to Banda Bahadur, a Sikh leader believed to have been executed in Mehrauli in the early 18th century.

This chapter began with a reading of the architecture of the structures within the Qutub complex, particularly urging a closer look not only at the inscriptions but also at the details of the forms and motifs used. Few studies, with the notable exception of that of Finbarr Flood,[25] have even begun to examine this aspect of these monuments. Most scholarly works still continue to base their conclusions on published official reports on the complex, even while they seek to read them critically. There has been little close study of the structures per se.

It is clear that an alternative reading is needed. The persistence of the interpretation of the complex as a symbol Muslim oppression of Hindus, casts its shadow nowhere more deeply than in its immediate vicinity, i.e. the village of Mehrauli. The colonial power in the 19th century did not merely reserve to itself the exclusive right to separate out and manage the complex as a monument. Through its official reports and guides, it put forth a canonical reading of the meaning of the structures, privileging its views over any others that might have existed. No attempt has been made to revise the official narrative in post-Independence India.

The Phoolwalon ki Sair festival, embodying pluralism under the Mughal polity, was prohibited by the British in 1942, at the height of the national movement. Revived in the 1960s with government support, it is still held every year. It bears a strongly official character, and its significance as a festival celebrating Hindu–Muslim amity *within the locality of Mehrauli* has been diluted by the participation of various delegations from all over the country. Its message, which could have resonated with a revised reading of the historic monuments of the Qutub, is lost in platitudes about religious amity. Against the dominant narrative centering on the Qutub monuments and the history of Mehrauli, it holds little significance for the local population.

Notes

1 For a history of the Ghurid conquest of North India, and particularly Delhi, see Sunil Kumar, *The Emergence of the Delhi Sultanate*, New Delhi: Permanent Black, 2007.
2 J. A. Page, *An Historical Memoir on the Qutb* (No. 22 of the Memoirs of the Archeological Survey of India), New Delhi: Swati Publications, 1991 (Reprint of 1926 edition), p. 29.
3 *Dilli Jain Directory*, New Delhi: Jain Sabha, 1961, p. 4.
4 Scholars have suggested that those temples that had been abandoned by their patrons before the conquest, or were not identified with political patrons, were left unharmed during conquest. See for instance Richard M. Eaton, 'Temple Desecration and the Indo-Muslim States', in Finbarr B. Flood (ed.), *Piety and Politics in the Early Indian Mosque*, New Delhi: Oxford University Press, 2008, pp. 69–70.
5 Mohammad Mujeeb, 'The Qutub Complex as a Social Document', in Flood (ed.), *Piety and Politics*, pp. 123–124.
6 Mujeeb, 'The Qutub Complex as a Social Document', pp. 128–129.
7 Page, *An Historical Memoir on the Qutb*, p. 43.
8 Muzaffar Alam, *The Languages of Political Islam: India, 1200–1800*, Chicago and London: University of Chicago Press, 2004, pp. 87–89.
9 Foreign Department, Miscellaneous Records, National Archives of India, New Delhi, (henceforth FMisc) vol 361, 6.1.51
10 Syed Ahmad Khan, *Asarussanadid*, New Delhi: Urdu Academy, 2000, p. 458.
11 Dehli Urdu Akhbar 13.9.1840; Dehli Urdu Akhbar 14.3.1841; FMisc vol 361, 25.5.1851
12 FMisc vol 361, 7.8.1851; 5–6,8–9.8.1852
13 Khan, *Asarussanadid*, p. 169; FP 25.4.1829 no. 16–19. Some of this construction can still be seen on the building.
14 Fanny Parks, *Wanderings of a Pilgrim in Search of the Picturesque*, Vol. II, London: Pelham Richardson, 1850, p. 205.
15 Foreign Department Proceedings, (Political), National Archives of India, New Delhi (henceforth FP) 25.4.1829 no. 20
16 M. M. Kaye (ed.), *The Golden Calm*, Exeter: Webb & Bower, 1980, pp. 146–148, 200.
17 FP 7.10.1848 no. 53
18 Khan, *Asarussanadid*, p. 294.
19 Sunil Kumar, *The Present in Delhi's Pasts*, New Delhi: Three Essays, 2002, pp. 7–10; Khan, *Asarussanadid*, pp. 166–168.
20 Mrinalini Rajgopalan, *Building Histories, the Archival and Affective Lives of Five Monuments in Modern Delhi*, Chicago and London: The University of Chicago Press, 2016, pp. 184–189.
21 FP 25.4.1829 no.20, 22; 4.9.1829 no.22; Khan, *Asarussanadid*, p. 169.
22 Khan, *Asarussanadid*, pp. 351–352.
23 Rajgopalan, *Building Histories*, pp. 176–177, 200–203.
24 *The Collected Works of Mahatma Gandhi*, Vol. 98, Ahemadabad: Publications Division, 1983, pp. 98–99; *Selected Works of Maulana Abul Kalam Azad*, Vol. 3, New Delhi: Atlantic Publishers and Distributors, 1991, p. 125.

25 Finbarr B. Flood, *Lost in Translation: Material Culture and Medieval "Hindu-Muslim" Encounter*, New Delhi: Permanent Black, 2009, p. 184.

Bibliography

Alam, Muzaffar, *The Languages of Political Islam: India, 1200–1800*, Chicago and London: University of Chicago Press, 2004.

Eaton, Richard M., 'Temple Desecration and the Indo-Muslim States', in Finbarr B. Flood (ed.), *Piety and Politics in the Early Indian Mosque*, New Delhi: Oxford University Press, 2008, pp. 69–70.

Flood, Finbarr B., *Lost in Translation: Material Culture and Medieval "Hindu-Muslim" Encounter*, New Delhi: Permanent Black, 2009.

Kaye, M. M. (ed.), *The Golden Calm*, Exeter: Webb & Bower, 1980.

Khan, Syed Ahmad, *Asarussanadid*, New Delhi: Urdu Academy, 2000.

Kumar, Sunil, *The Present in Delhi's Pasts*, New Delhi: Three Essays, 2002.

Kumar, Sunil, *The Emergence of the Delhi Sultanate*, New Delhi: Permanent Black, 2007.

Mujeeb, Mohammad, 'The Qutub Complex as a Social Document', in Finbarr B. Flood (ed.), *Piety and Politics in the Early Indian Mosque*, New Delhi: Oxford University Press, 2008.

Page, J. A., *An Historical Memoir on the Qutb* (No. 22 of the Memoirs of the Archeological Survey of India), New Delhi: Swati Publications, 1991 (Reprint of 1926 edition).

Parks, Fanny, *Wanderings of a Pilgrim in Search of the Picturesque*, Vol. II, London: Pelham Richardson, 1850.

Rajgopalan, Mrinalini, *Building Histories, the Archival and Affective Lives of Five Monuments in Modern Delhi*, Chicago and London: University of Chicago Press, 2016.

Part III
Transnational heritage

8 Beyond World Heritage

Lumbinī – the creation of a more meaningful site?

Max Deeg

This chapter analyses the interpretation of the archaeological and historical evidence of the birthplace of the Buddha, Lumbinī (Terai, Nepal), since its inscription as World Heritage Site in 1997 as one of Nepal's four sites on the World Heritage List.[1] It will trace how the structure of meaning of the site has changed after its recognition as World Heritage Site through archaeological activities (conservation) and reconstruction of history (interpretation) in the framework of particular ideological parameters. The chapter will claim that in the case of Lumbinī this happened through a series of bold and distorted conclusions from archaeological excavations, both in the dimensions of space (where?) and time (when?). One of these conclusions is based on the "discovery" of the so-called Marker Stone in the lower strata underneath the former Māyā-devī temple which has been interpreted as marking the exact spot where Siddhārtha was born. The stone became the centre of the newly built structure above the archaeological site – in a way shifting the centre of the site from the Ashokan pillar – and soon attracted, through its postulated meaning, the attention of Buddhist pilgrims. Another, even more recent, claim that would make Lumbinī the oldest identifiable religious shrine in South Asia is that on its lowest stratum a pre-Buddhist sanctuary has been found. These two examples raise a number of questions such as: Is the rationale for the inscription as a heritage site a final definition? To what extent is it reasonable to "re-read" and "re-write" a heritage site after inscription? Is such a reinterpretation of the site in a wider public space, on the basis of new material or historical evidence (real or claimed), feasible and in accordance with the concept of heritage? How do these reinterpretations influence the public "usage" and notion of the site? The chapter will address these issues and the role of scholarly discourse in relation to other agenda such as the ideologies of nation state or religious identities.

When the Nepalese general Khadga Shamsher started excavating a site in the Nepalese Terai near the Nepalese–British border in 1896 no one could foresee that this place would become, a little bit more than 100 years later, one of the recognised places of World Heritage, or 'one of the most sacred places of the world'.[2]

The last statement is certainly arguable from different standpoints, not least from an emic Buddhist one. Although the often-quoted reference of the Buddha before his physical death or *parinirvāṇa* to the four major places of pilgrimage in the Mahāparinirvānasūtra (P. Mahāparinibbānasuttanta, Dīghanikāya) clearly highlights the importance of the birthplace of the Buddha, the historical and archaeological evidence points in a slightly different direction, with Bodhgaya, the site of the Buddha's full Enlightenment (*samyaksaṃbodhi*), and Sarnath, the place of the First Sermon (*dharmacakrapravartana*, 'Setting in Motion the Wheel of the Dharma') of the Enlightened One, being much more prominent in textual and material sources. Yet the former Secretary General of the United Nations, U Thant, himself a Burmese Buddhist, after a visit to the place in 1967 selected Lumbinī to become developed as a Buddhist place of importance and pilgrimage and finally inscribed as World Heritage.

The so-called nativity has never been, in Buddhist history and view, the most important site in terms of soteriological meaning compared with the places of the Enlightenment, First Sermon or *Parinirvāṇa* (the Great Extinction). The place was clearly upgraded after U Thant's visit and his plea to the United Nations in 1970 to develop Lumbinī. The establishment of this newly ascribed importance was influenced, one could argue, by a Christo-centric tendency, looking, as it were, for the Buddhist "Bethlehem" – which had already been reflected in the British archaeologists' hunt for the birthplace of the Buddha (and his hometown Kapilavastu) in the second half of the 19th century.[3] This hunt went hand in hand with the historicisation of the Buddha and his biography through Western Buddhist scholars. This, on the other hand, had a clear impact on the way the Pan-Buddhist movement under the leadership of people like Anāgārika Dharmapāla under the emerging organisational roof of the Mahābodhi Society acted when they claimed back, for instance, the sacred sites like Bodhgaya or Sarnath linked to the major events in the historical life of the Buddha.

Lumbinī, as the birthplace of the Buddha was unanimously called in Buddhist biographical texts such as the *Lalitavistara* and the *Mahāvastu* and other sources, was the last of the major Buddhist sacred places to be discovered. Alexander Cunningham, the first director of the Archaeological Survey of India and "founder" of a kind of systematised Indian or South Asian archaeology, had been puzzled by the riddle of the birthplace's identity and location, which he had been unable to solve during his attempts to work out the relation between Chinese travelogues and archaeological sites[4] until he finally left India in 1885.[5] The centre of Lumbinī[6] as a place of memory – in spatial terms rather remembering than constituting a continuity of memory since knowledge about the place had been lost until its (re-)discovery at the end of the 19th century – has been and is clearly the Ashokan pillar.[7] Only through its inscription was the identification of the place as the birthplace of the Buddha made possible and plausible.[8] Already in 1898–1899 Georg Bühler, who made one of the first readings and translations of the inscription after its discovery, stated:

> The greatest importance of the inscription . . . for the topography of ancient India and the sacred history of Buddhism has been fully recognised by Dr. Führer. . . . It fixes with absolute certainty the situation of the garden of Lumbinī where according to the Buddhist tradition prince Siddhârtha was born.[9]

The pillar inscription, with caveats and one or two words disputed and unclear, can be translated as:

> King Priyadarśin, who is dear to the gods, came here in the twentieth year following his consecration and paid reverence. Thinking (*iti*), 'Here the Buddha was born – the muni of the Śākya clan', he caused a *vigaḍabhīcā* of stone to be made and a pillar of stone to be erected. Thinking, "Here the Lord was born", he exempted the village of Lumbinī from imposts and had it receive the "eight rights".[10]

The pillar had been linked with the place of the Buddha's birth from an early time onwards. While the earliest Chinese record by the Buddhist traveller Faxian 法顯 from the beginning of the 5th century CE does not, strangely enough, mention an Ashokan pillar *in situ*, Xuanzang 玄奘 in the first half of the 7th century clearly refers to this most striking feature of the place:

> Not far away from the *stūpa* of the four heavenly kings carrying the prince there was a big stone pillar erected by king Aśoka with a horse statue on top [of it]. Later it broke apart in the middle through the thunder [strike] of an evil *nāga*.[11]

Since the re-discovery of Lumbinī and, until recently, the pillar (the statue on top has not been preserved except for some small fragments) has not only been the centre of attention by scholars working on Lumbinī but also has been, together with the huge pipal tree south of it and on the opposite side of the water tank, the focus of pilgrims' and Buddhists' worship.[12] This situation changed at the beginning of the present millennium with the propagandistic approach to the results of a round of archaeological excavations by the archaeologists involved and the Lumbinī Development Trust (LDT).[13]

In 1997, the year of the inscription of Lumbinī into the list of World Heritage monuments, the former so-called Māyādevī temple had already been dismantled, and the site, being partly and slowly (with interruptions) excavated between 1992 and 1996,[14] was for a while an exposed heap of bricks barely protected by sheets of plastic. After the excavation a new "Māyādevī temple" in form of a plain rectangular and later white-washed brick building with quasi-Tibetan stylistic elements was erected over the site to "protect" the excavated and exposed oldest layers of the structure beneath the old Māyādevī sanctuary. This new structure has been a constant object of critique in UNESCO reports as not being suitable to protect the excavated ruins in terms of form and functionality.

In 1994 the so-called Marker Stone[15] was discovered during the excavations undertaken by the Japanese Buddhist Federation (JBF) under the supervision of the archaeologist *in situ*, Satoru Uesaka, and the Nepalese Department of Archaeology.[16] What is described as a "conglomerate stone" with an alleged origin from the Śivalikh range north of the Terai plain was bound to puzzle but also to fire the imagination of the Japanese and Nepalese archaeologists who discovered the object. In the 2001 excavation report[17] of the Japanese, which is full of linguistic and factual mistakes, this becomes quite obvious. The report first describes the excavated structures, the cells or chambers of what has been explained as a "sanctuary" or "temple", in a standard way. In the description of the so-called Chamber II the Marker Stone is introduced the first time without any further discussion or explanation why it is so called: 'Under the upper structure, a thin Marker Stone in north-south direction was settled on the center part. . . . It is considered that the destruction at stage II was to observe this marker stone'.[1]Japan Buddhist Federation, 'Archaeological Research', p. 50.

At the end of the description we find a brief discussion of the conglomerate stone with the remarks of an "expert":

Expert Opinion on the So-called Conglomerate

1 The surface seems to weather.
2 It is comparatively soft.

3 It is not conglomerate but a mass of gravel bound with lime.
4 it is artificial.[18]

On p. 64, however, the report suddenly introduces the "Marker Stone" and its function or meaning by stating (emphasis added):

> Facts about stage I [the earliest period of construction] are:
>
> a fifteen chambers regularly arranged, rectangle in the longitu-dinal direction of east–west, and double-walled;
> b Marker Stone presumed to have been set up in Chamber 2 *to show Buddha's nativity point*;
> c charcoal found beneath the foundation plane of chamber 8.[19]

A final remark on the original chamber structure which already had been excavated emphasises the centrality of what is by then established to be the Marker Stone of the Buddha's actual birthplace:

> Structures vary in shape through stages. However, they are con-sistent to center on the marker stone that was set up in chamber 2 at stage I.[20]

What follows is a bold interpretation of the Marker Stone without any substantial internal or external support of evidence. After a descrip-tion of the excavation situation of the stone the author(s) abruptly jump(s) to a bold conclusion: 'People might have known of the nativity point when they began to built [sic!] the Maya Devi Temple, and after-wards they heard the tradition in detail. Then they might have decided to instal a stone to mark the very point'.[21] Then – in a kind of herme-neutical circle – an expert's opinion by Professor A. K. Narain is quoted at length to "support" the original interpretation. Narain suggests that the stone is linked to the mysterious *hapax legomenon* *sīlavigaḍabhīcā* of the Ashokan inscription – which he, unfortunately, does not expain philologically, claiming that '[t]he historical Buddha, Siddhārtha Gau-tama, the Shakyamuni, has become more historical now with the dis-covery . . . of an enshrined piece of Stone Marker'.[22]

The sheer fantastic nature of this conclusion, repeated several times, and Narain's suggestion is demonstrated in the following passage:

> Obviously this stone marker was supposed to indicate the exact place made sacred by the birth of Shakyamuni Buddha. . . . Unfor-tunately for us, this stone does not bear any inscription recording the event and identifying the purpose of its placement and that of the associated structure around it. But this need not surprise us

irrespective of either of the two possibilities we may reasonably envisage. The first is that this marker was placed by the Sākyas themselves long before the time of Asoka, who is believed to have been the first to introduce the Brāhmi script of at least standing the practice of writing or engraving it on stone. The second possibility is that the fact of the brick structure enclosing the stone marker and its purpose is after all recorded in the Brāhmi script for the first time by Asoka on the stone pillar (*Silāthabha*) setting up (*Usapāpite*) by him at Lumbini. Also that, it was Asoka himself who, in addition to the setting up (*Usapāpite*) of the stone pillar (*Silāthabhe*), caused (*kālāpita*) the building of the brick structure with the extraordinary stone referred to as *Silāvigadabhicā*. Whichever possibility is accepted, we believe that, as shown below, the baffling mystery of a written reference to it in the text of the Rummindei Pillar inscription of Asoka now stands resolved.[23]

The text then goes into a not less phantasy-driven philological explanation of the *hapax silāvigaḍabhīcā*, identifying it finally as a word for the Marker Stone in the sense of 'the much spoken of (unusual) stone'.[24]

Without going into a detailed refutation of what is not even a sound line of argument, it may suffice to point out some points which could have been discussed and brought up critically in the context of the so-called Marker Stone and its meaning:

1   Why should an unembellished and rather vulgar piece of stone of a completely irregular shape have been used to mark the exact *locus* of what must have been an extremely important place, the exact spot of the birth of the Buddha, for the Buddhist tradition?
2   Why is the stone not placed in the very centre of the Mauryan structure if it marked such an important spot?
3   Where is the reflection of this important object in Buddhist biographical literature which originates from times of which we know that Lumbinī as a pilgrimage place was still in use?

Over the next couple of years the stone was transformed into and became an uncritically accepted meaningful relic, as can be seen in the following sentence in Weise's documentation of Lumbinī as a heritage site: 'The most important finding from their [the Japanese's] excavation is the Marker Stone, which has been interpreted by many scholars as representing the exact birth place of Lord Buddha'.[25]

A statements like this is, as is immediately evident, problematic in several respects. First, it assumes a kind of final authority by the implied claim that the so-called Marker Stone was the most important finding of the Japanese archaeological campaign. But it also pretends that this is a generally accepted scholarly view ('which has been interpreted by many scholars as representing') while in reality no discussion was possible on the issue until – theoretically at least – the delayed publication of the Japanese excavation reports from 2001 and later.

In the meantime, however, the interpretation of the Marker Stone had made its imprint on the site. Although the stone is never mentioned in any document of the UNESCO heritage committee in the process of the inscription[26] it became central in the layout of the newly erected building on top of the exposed excavation site, sometimes called the new Māyādevī temple, and therefore became also the focus and centre of veneration by Buddhist monastics and pilgrims. The Ashokan pillar which had been, together with the huge bodhi tree south of it, the centre of pilgrims' activities and veneration and also ranked prominently in the UNESCO documentation at the time of Lumbinī's inscription became marginalised in spatial and symbolic terms. In order to demonstrate how this shift happened I quote passages from the description of the site on the UNESCO homepage in which the Marker Stone is not mentioned at all (emphasis added):

Lumbini, the birthplace of the Lord Buddha

Siddhartha Gautama, the Lord Buddha, was born in 623 BCE in the famous gardens of Lumbini, which soon became a place of pilgrimage. *Among the pilgrims was the Indian emperor Ashoka, who erected one of his commemorative pillars there.* The site is now being developed as a Buddhist pilgrimage centre, where the archaeological remains associated with the birth of the Lord Buddha form a central feature.

Outstanding universal value

Brief synthesis

The Lord Buddha was born in 623 BCE in the sacred area of Lumbini located in the Terai plains of southern Nepal, *testified by the inscription on the pillar erected by the Mauryan Emperor*

Asoka in 249 BCE. Lumbini is one of the holiest places of one of the world's great religions, and its remains contain important evidence about the nature of Buddhist pilgrimage centres from as early as the third century BCE.

The complex of structures within the archaeological conservation area includes the Shakya Tank; the remains within the Maya Devi Temple consisting of brick structures in a cross-wall system dating from the third century BCE to the present century and *the sandstone Ashoka pillar* with its Pali inscription in Brahmi script. Additionally there are the excavated remains of Buddhist viharas (monasteries) of the third century BCE to the fifth century CE and the remains of Buddhist stupas (memorial shrines) from the third century BCE to the 15th century CE. The site is now being developed as a Buddhist pilgrimage centre, where the archaeological remains associated with the birth of the Lord Buddha form a central feature.

Criterion (iii): As the birthplace of the Lord Buddha, testified by *the inscription on the Asoka pillar*, the sacred area in Lumbini is one of the most holy and significant places for one of the world's great religions.

Criterion (vi): The archaeological remains of the Buddhist viharas (monasteries) and stupas (memorial shrines) from the third century BCE to the 15th century CE, provide important evidence about the nature of Buddhist pilgrimage centres from a very early period.

Integrity

The integrity of Lumbini has been achieved by means of preserving the archaeological remains within the property boundary that give the property its Outstanding Universal Value. The significant attributes and elements of the property have been preserved. The buffer zone gives the property a further layer of protection. Further excavations of potential archaeological sites and appropriate protection of the archaeological remains are a high priority for the integrity of the property. The property boundary however does not include the entire archaeological site and various parts are found in the buffer zone. The entire property including the buffer zone is owned by the Government of Nepal and is being managed by the Lumbini Development Trust and therefore there is little threat of development or neglect. However the effects of industrial development in the region have been identified as a threat to the integrity of the property.

Authenticity

The authenticity of the archaeological remains within the boundaries has been confirmed through a series of excavations since the discovery of *the Asoka pillar* in 1896. The remains of viharas, stupas and numerous layers of brick structures from the third century BCE to the present century at the site of the Maya Devi Temple are proof of Lumbini having been a centre of pilgrimage from early times. The archaeological remains require active conservation and monitoring to ensure that the impact of natural degradation, influence of humidity and the impact of the visitors are kept under control. The property continues to express its Outstanding Universal Value through its archaeological remains. The delicate balance must be maintained between conserving the archaeological vestiges of the property while providing for the pilgrims.

Protection and management requirements

The property site is protected by the Ancient Monument Preservation Act 1956. The site management is carried out by the Lumbini Development Trust, an autonomous and non-profit making organisation. The entire property is owned by the Government of Nepal. The property falls within the centre of the Master Plan area, the planning of which was initiated together with the United Nations and carried out by Prof. Kenzo Tange between 1972 and 1978.

The long-term challenges for the protection and management of the property are to control the impact of visitors, and natural impacts including humidity and the industrial development in the region. A Management Plan is in the process of being developed to ensure the long-term safeguarding of the archaeological vestiges of the property while allowing for the property to continue being visited by pilgrims and tourists from around the world.[27]

The UNESCO has, since the inscription of Lumbinī in December 1997 and starting in 1999, issued 14 "State of Conservation" reports[28] on Lumbinī. In these reports the Marker Stone gradually slips in as an important archaeological and historical item. As far as I can see it is in the 2004 report ('Conservation issues presented to the World Heritage Committee') that the Marker Stone is mentioned the first time in a UNESCO document (emphasis added):

The report submitted by the State Party to the Secretariat on 21 January 2004 draws attention to the conflicting recommendations made by successive expert missions on whether the excavated part of the shrine should be covered or left exposed. It also mentions that the Maya Devi shrine is a living pilgrimage site with worldwide spiritual value, incessantly visited by devotees, whose religious sensibilities had been affected by the disagreeable conditions whereby the *Marker Stone and the image of Maya Devi* were accommodated in a temporary shelter. It was due to this situation that the decision was taken to reinstall a permanent protective structure.[29]

The report also indicates the provision of a management plan of the holy complex which includes issues regarding entry regulation, guidance and facilities for visitors, promotional activities, access control to the main sanctum, logging of weather record and security measures of the property. It also highlights the action taken by the State Party, which fully implemented the UNESCO recommendations made at the 2001 International Scientific Experts Meeting, using them as guidelines for the restoration work of the Maya Devi shrine.

A UNESCO/International Council on Monuments and Sites (ICOMOS) joint mission undertaken in May 2004 assessed the impact of the newly constructed Maya Devi shrine on the World Heritage value of the property as a whole. The mission findings and recommendations will be reported at the time of the 28th session of the Committee in 2004.

The report from the year 2006, while still referring to the stone by inverted commas, refers again to the Marker Stone as the essential feature of the site to be conserved:

> ICOMOS . . . draws the attention of the Committee to the following additional points, raised in the World Heritage Centre–ICOMOS mission report, but not directly addressed in the report's recommendations:
>
> a) *The fragility of the so-called Marker Stone*, maintained in situ under the new temples in ground water conditions which keep it constantly wet; . . . [30]

The Marker Stone has, in a period of 15 years, become the centre of the Lumbinī site, marginalising the former three focal elements – Ashokan pillar, pipal tree and pond – as can be seen in the "Report on the State of Conservation of the property" from November 2015, submitted to the UNESCO by the Nepalese Department of Archaeology. In the list of 'Major Buddhist Sites in the Greater Lumbini Area', beside other interpretational plunder,[31] Lumbinī is described as 'Birthplace of Buddha with Marker Stone'.[32] In the report of the years 2010–2013, 'Strengthening Conservation and Management of Lumbini, the Birthplace of the Lord Buddha, World Heritage Property: A UNESCO/Japanese Funds-in-Trust Project' (http://whc.unesco.org/en/activities/729/, accessed 30 June 2016), the Marker Stone has risen to one of the three major features of the site the conservation of which is given priority:

Objectives & activities

1   Resolution of conservation problems affecting the [*Asoka Pillar, the Marker Stone, and the Nativity Sculpture*] and implementation of a monitoring system for the Maya Devi Temple;
2   Improvement of knowledge of the archaeological vestiges contained within the boundaries of the World Heritage property, particularly in the buffer zone, in order to proceed with the planning of the Sacred Garden;
3   Review of the present state of the Sacred Garden vis-à-vis the Kenzo Tange Master Plan, particularly the current abstract grid pattern and the drainage and water levels of bodies of water around the Sacred Garden, and preparation of an Interim Plan based on the project activities, including a moratorium on certain activities for the entire Sacred Garden area as defined in the Master Plan;
4   Finalisation, adoption and implementation of the Integrated Management Plan (IMP) in order to preserve the Outstanding Universal Value of Lumbini in the long term;
5   Capacity building for Nepalese experts, especially the conservation and archaeological staff of the Lumbini Development Trust and the Department of Archaeology;

Establishment of scientific documentation for the above activities.[33]

As can be seen, the UNESCO documentation of Lumbinī is closely following the established narrative of the central meaning of the Marker Stone for the whole site. What is worse from a scholarly standpoint is that the Marker Stone – I deliberately refrain from using the capitalised Marker Stone – instead of its meaning being critically questioned, has been accepted as a key point in the archaeological stratification and dating process. This is shown in a recent publication by Durham archaeologist Robin Coningham and others on their latest excavations at the Lumbinī site, with results that raised the critical eyebrows of "traditional" Buddhologists.[34] The article takes the earlier interpretation of the Marker Stone for granted: 'This later phase [of the Māyādevī Temple area] drastically altered the Temple layout and focus of veneration with the installation of the "Marker Stone"'.[35] With such a statement the uncritical acceptance of the stone and its meaning has been codified and the claim of it marking a centre of veneration at the time of Ashoka is now supported "scientifically". Fortunately this sequence of scholarly discourse following and serving popular institutionalisation does not happen very often.

A neutral and critical review of the situation in the light of the historical and archaeological evidence has to leave the observer rather puzzled how an object the significance of which is completely unclear could have become loaded up with such meaning as representing the exact birthplace of the Buddha by the random interpretation by less than a handful of archaeologists in the form of a rather doubtful and dubious historical conclusion.

In my view, construction of the meaning of the Marker Stone at Lumbinī is a striking example of irresponsibly uncritical scholarship translating its highly questionable conclusions not only into popular notions and assumptions about the history and meaning of particular sites but also into practically irreversible changes of religious practice. The process of interpretation and ascription of meaning to the stone is almost like the one Richard Gombrich has described with respect to Durham archaeologist Robin Coningham's claim to have solved the problem of the date of the Buddha's birth by archaeological evidence at Lumbinī briefly discussed earlier:

> as hypothesis, or rather, guess, builds upon guess, possible slides into probable and finally emerges unembellished as firm claim.[36]

What looks like a minor case of interpretation of one archaeological object in Lumbinī, the Marker Stone, raises some questions related to the treatment of heritage (and not necessarily restricted to Word Heritage) post declaration or inscription/recognition: Is it justifiable to increase, or create even more, "Outstanding Universal Value"[37] by a reinterpretation of sites which is not in accordance with standard scholarly and scientific standards? Is the rationale underlying the inscription as a heritage site a final definition of a site's historical value? To what extent, under which circumstances, and with what methods is it reasonable and acceptable to "re-read" and "re-write" a heritage site after inscription? Is such a reinterpretation of the site in a wider public space, on the basis of new material or historical evidence, especially in case of highly questionable reinterpretations of the evidence, feasible and in accordance with the concept of heritage? Who has the supremacy of interpretation of a site and its meaning? How do these reinterpretations influence the public "usage" and notion, national and international, of the site?

This is certainly not the place to answer all these questions. The purpose of this chapter is rather to open a discussion on these issues that the example of Lumbinī and the "Marker Stone" may illustrate. However, I am sure that similar problems are connected with other heritage sites, not only in South Asia but also elsewhere. The basis for such a discussion should be the criteria of authenticity and integrity fleshed out in the 'Operational Guidelines for the Implementation of the World Heritage Convention' which Brigitta Ringbeck has paraphrased:

> Authenticity refers to the truthful and credible conveyance of the historical and cultural significance of the site. Depending on the cultural context, authenticity has to be expressed in a convincing and genuine matter through numerous matters. Authenticity manifests itself in form and composition, material and substance, use and function, techniques and administrative systems, location and overall context and other expressions. Therefore a site must express a multidimensional meaning and symbolism attested by scientific research. . . . Integrity refers to the wholeness of a World Heritage site. With regard to cultural heritage the physical substance should be in good, conservationally controlled condition. The preservation of visual integrity is also decisive; this affects the overall aesthetic impression of a site, its unhindered perceptibility and its dominating effect from a distance.[38]

In the light of the definition of these two crucial criteria for a World Heritage Site – but I would claim, also for any other heritage site – one may ask the question whether the shift of focus from the Ashokan pillar to the "Marker Stone", hidden away in the newly constructed building in Lumbinī, has contributed anything meaningful to the site as a whole. One could argue that the authenticity of the new object in the centre is questionable, and that the new focus on it is disturbing the integrity of the site. While attempts at improving the historical value of a heritage site by procuring more specific data and/or artefacts endorsing the meaning by being closer or more directly linked to the events ascribed to the place by a cultural or religious tradition are understandable, these aims should not dominate the scholarly interpretation of such data or objects in the first place.

The broader question of power of interpretation in relation to sites raised in the definition of authenticity cited earlier and at the same time referring to the notion of contested cultural heritage in a wider sense[39] is not less important than the question of material ownership.[40] The construction of meaning at an archaeological heritage site like Lumbinī demonstrates the complex entanglement of several stakeholders such as local societies, religious groups, national politics, international politics, and academic disciplines like archaeology and history. The example can show how a restricted interpretation of a particular group of stakeholders is able to influence the international perception of a site and its historical meaning apart from and beyond a wider public and academic discourse needed to fulfil the criteria of authenticity and integrity.

Notes

1 The three other inscribed Nepalese World Heritage sites are the Kathmandu Valley, the Sagarmatha (Mount Everest) National Park and the Chitwan National Park.

2 K. Weise (ed.), *The Sacred Garden of Lumbini: Perceptions of the Buddha's Birthplace*, Paris: UNESCO, 2013, p. 5. This is obviously a rephrasing and re-shifting of what is found in the World Heritage Committee's recommendation as 'one of the holiest places of one of the world's great religions' (quoted in Weise, *The Sacred Garden of Lumbini*, p. 127).

3 H. Falk, *The Discovery of Lumbinī*, Lumbini: Lumbini International Research Institute, 1998.

4 See, for example, A. Cunningham, 'Verification of the Itinerary of the Chinese Pilgrim, Hwan Thsang, Through Afghanistan and India, During the First Half of the Seventh Century of the Christian Era', *The Journal of the Asiatic Society of Bengal*, 1848, 17(2): 29, where he identifies in a combination of directions and distances given by the Chinese travellers and fanciful etymologies of place names Lumbinī with Jaunpur. Cunningham changed his mind about the location of Lumbinī several times and was harshly criticised for his mis-identifications by contemporary scholars.

5 A. Imam, *Sir Alexander Cunningham and the Beginnings of Indian Archaeology*, Dacca: Asiatic Society of Pakistan, 1966, p. 168f; Falk, *The Discovery of Lumbinī*, p. 8f.

6 On the history of the site see M. Deeg, *The Places Where Siddhārtha Trod: Lumbinī and Kapilavastu*, Lumbini: The Lumbinī Internation Research Institute, 2004. On the various aspects of Lumbinī in Buddhist texts, archaeological context, Buddhist pilgrimage and practice see the various articles collected in C. Cueppers, M. Deeg and H. Durt (eds.), *The*

Birth of the Buddha: Proceedings of the International Conference Held in Lumbini, Nepal, October 2004, Lumbini: Lumbini International Research Institute, 2010.

7 H. Falk, *Aśokan Sites and Artefacts – a Source Book With Bibliography*, Mainz: Philip von Zabern, 2006, pp. 177–180.

8 Attempts at discrediting this identity as the one by the British Terence Phelps on his website (www.lumkap.org.uk [accessed on 30 June 2016]), who in 2009 transformed the content of his website into a book (T. Phelps, *The Lumbini and Piprahwa Deception*, London: Terence Phelps Ltd., 2009), can be neglected. The acceptance of Phelps' denial of Lumbinī in Nepal being the Buddha's birthplace, however, can lead to utterly absurd speculations, as can be seen, for instance, in R. Pal's, 'The Dawn of Religion in Afghanistan-Seistan-Gandhara and the Personal Seals of Gotama Buddha and Zoroaster', *Mithras Reader – an Academic and Religious Journal of Greek, Roman and Persian Studies*, 2010, 3: 62–83, who tries to prove that the Buddha was Persian.

9 G. Bühler, 'The Asoka Edicts of Paḍêria and Nigliva', *Epigraphia Indica*, 1898–99, 5(2).

10 *Devānapiyena piyadasina lājina vīsativasābhisitena atana āgāca mahīyite hida budhe jāte sakyamunī ti silāvigaḍabhī cā kālāpita silāthabhe ca usapāpite hida bhagavaṃ jāte ti luṃminigāme ubalike kaṭe aṭhabhāgiye ca*. I have adapted the translation of Falk, *The Discovery of Lumbinī*, p. 16; see also Falk, *Aśokan Sites and Artefacts*, p. 180. For the various interpretations of the two cruces in the inscription, vigaḍabhīcā and aṭhabhagiya, see Falk, *Aśokan Sites and Artefacts*, p. 179. On vigaḍabhīcā more specifically see H. Falk, 'The Fate of Aśoka's Donations at Lumbinī', in P. Olivelle, J. Leoshko and H. P. Ray (eds), *Reimagining Aśoka – Memory and History*, Oxford: Oxford University Press, 2012, pp. 204–216, who believes that the term refers to stone railing around a platform a part of which is preserved at Sārnāth where a part of the relics of the Buddha donated to Lumbinī by Aśoka was stored.

11 Datang xiyu ji 大唐西域記 ('Record of the Western Regions of the Great Tang [Dynasty]'), Taishō-shinshū-daizōkyō (T.) 2087.902b.12ff.: 四天王捧太子窣堵波側不遠，有大石柱，上作馬像，無憂王之所建也。後為惡龍霹靂，其柱中折仆地。; the translation is my own: see Deeg, *The Places Where Siddhārtha Trod*, p. 55.

12 In the Japanese 2001 excavation report the Pakistani scholar Hassan Ahmad Dani (contribution dated 17 March 1995), despite the publication's attempt to already draw the focus on the recently discovered 'marker stone', still suggested to make the so-called nativity sculpture the centre of pilgrims' veneration (Japan Buddhist Federation, *Archaeological Research at Mayadevi Temple*, Lumbini, Tokyo: Japanese Buddhist Federation, 2001, p. 34).

13 This change is reflected in the survey on the activities at the Sacred Garden published by R. A. E. Coningham, A. Schmidt and K. M. S. Strickland, 'A Cultural and Environmental Monitoring of the UNESCO World Heritage Site of Lumbini, Nepal', *Ancient Nepal*, March 2011, p. 176, particularly p. 7, who give a the focus of visitors on the Maya Devi temple (with the marker stone inside) as 44% and on the Aśokan pillar as 15.3%. The authors themselves – consciously or unconsciously – bought into the

importance and centrality of the marker stone when they took climatic measurements (rainfall, temperature, humidity) 'when possible, close to the Marker Stone chamber' (p. 1).

14 Weise, *The Sacred Garden of Lumbini*, p. 73.

15 A good photograph of the stone is found in R. A. E. Coningham, K. P. Acharya, K. M. Strickland, C. E. Davis, M. J. Manuel, I. A. Simpson, K. Gilliland, J. Tremblay, T. C. Kinnaird and D. C. W. Sanderson, 'The Earliest Buddhist Shrine: Excavating the Birthplace of the Buddha, Lumbini (Nepal)', *Antiquity*, 2013, 87: 2013: 1110, Figure 5.

16 Weise, *The Sacred Garden of Lumbini*, p. 48.

17 The report from 2003 (Japan Buddhist Federation, *Lumbini: The Archaeological Survey Report 1992–1995*, Tokyo: Japan Buddhist Federation, 2003), being more or less an abridged repetition of the 2001 report, does not contain new information about the stone.

18 Japan Buddhist Federation, *Archaeological Research*, p. 56. The judgement is repeated and the experts are later (p. 77) identified as a Dr. Koshiro Kizaki and Dr. B. N.Upreti, Tribhuvan University, Department of Geology.

19 Japan Buddhist Federation, *Archaeological Research*, p. 64.

20 Japan Buddhist Federation, *Archaeological Research*, p. 66.

21 Japan Buddhist Federation, *Archaeological Research*, p. 67.

22 Japan Buddhist Federation, *Archaeological Research*, p. 68.

23 Japan Buddhist Federation, *Archaeological Research*, p. 71.

24 Japan Buddhist Federation, *Archaeological Research*, p. 75. It is not clear who the scholar behind the 'I' of the text is.

25 Weise, *The Sacred Garden of Lumbini*, p. 73. Although the different parts and chapters of the book are not marked as individually authored it is very probable that this statement comes from Coningham. See also Weise, *The Sacred Garden of Lumbini*, p. 84.

26 Weise, *The Sacred Garden of Lumbini*, p. 131ff.

27 UNESCO description, http://whc.unesco.org/en/list/666 (accessed 30 June 2016).

28 http://whc.unesco.org/en/soc/2346 (accessed 30 June 2016); almost annually except 2007, 2010, 2013, 2015.

29 http://whc.unesco.org/en/soc/1460 (accessed 30 June 2016)

30 http://whc.unesco.org/en/soc/1179, (accessed on 30 June 2016).

31 See 'Niglihawa: Birthplace of Kanakamuni Buddha', 'Araurakot: Hometown of Kanakamuni Buddha', 'Sagarhawa: Ruined stupa for King Mahanama – a martyr for people – built by King Asoka', 'Ramgram: The only Stupa having the remains of Buddha intact'.

32 Government of Nepal, Ministry of Culture, Tourism and Civil Aviation, Department of Archaeology, Lumbini, the Birthplace of the Lord Buddha (Nepal) (C 666rev). A report on the State of Conservation of the property, Kathmandu, 2015, p. 10.

33 http://whc.unesco.org/en/activities/729/

34 Coningham et al., 'The Earliest Buddhist Shrine', p. 1107. See also the marker stone marked red on the plan of the Maya Devi Temple (1109, Figure 4, section C2). One example of harsh criticism of Coningham's conclusions is a piece written by the former Bowden chair in Sanskrit at Oxford University Richard Gombrich ('Pseudo-Discoveries at Lumbini', 27

November 2013, http://ocbs.org/wp-content/uploads/2015/09/rfg1213. pdf [accessed on 16 July 2016]).
35 Coningham et al., 'The Earliest Buddhist Shrine', p. 1110.
36 To push Gombrich's remarks even further and into the more concrete context of South Indian archaeology and history one could say that this procedure is following the 'Cunninghamian' method of identifying meaningful places and sites on the basis of textual sources (mainly the Chinese Buddhist records) with a reluctance to use these sources in a proper and contextualized way.
37 See S. Labadi, *UNESCO, Cultural Heritage, and Outstanding Universal Value: Value-Based Analyses of the World Heritage and Intangible Cultural Heritage Conventions*, Lanham and New York: AltaMira Press, 2013.
38 B. Ringbeck, *Management Plans for World Heritage Sites: A Practical Guide*, Bonn: German Commission for UNESCO, 2008, p. 17f. The original formulation of the Guidelines is found on p. 91ff.
39 See H. Silverman, *Contested Cultural Heritage: Religion, Nationalism, Erasure, and Exclusion in a Global World*, New York and Dordrecht, The Netherlands: Springer, 2011.
40 See J. Cuno, *Who Owns Antiquity? Museums and the Battle Over Our Ancient Heritage*, Princeton, NJ and Oxford: Princeton University Press, 2008.

Bibliography

Allen, Charles, *The Buddha and Dr Führer: An Archaeological Scandal*, London: House Publishing, 2011.
Bühler, Georg, 'The Asoka Edicts of Paḍêria and Nigliva', *Epigraphia Indica*, 1898–99, 5: 1–6.
Coningham, Robin A. E., K. P. Acharya, K. M. S. Strickland, C. E. Davis, M. J. Manuel, I. A. Simpson, K. Gilliland, J. Tremblay, T. C. Kinnaird, and D. C. W. Sanderson, 'The Earliest Buddhist Shrine: Excavating the Birthplace of the Buddha, Lumbini (Nepal)', *Antiquity*, 2013, 87: 1104–1123.
Coningham, Robin A. E., A. Schmidt, and K. M. S. Strickland, 'A Cultural and Environmental Monitoring of the UNESCO World Heritage Site of Lumbini, Nepal', *Ancient Nepal*, March 2011, 176: 1–8.
Cueppers, Christoph, Max Deeg, and Hubert Durt (eds), *The Birth of the Buddha: Proceedings of the International Conference Held in Lumbini, Nepal, October 2004*, Lumbini: Lumbini International Research Institute, 2010.
Cunnningham, Alexander, 'Verification of the Itinerary of the Chinese Pilgrim, Hwan Thsang, Through Afghanistan and India, During the First Half of the Seventh Century of the Christian Era', *The Journal of the Asiatic Society of Bengal*, 1848, 17(2): 13–60.
Cuno, James, *Who Owns Antiquity? Museums and the Battle Over Our Ancient Heritage*, Princeton, NJ and Oxford: Princeton University Press, 2008.

Deeg, Max, *The Places Where Siddhārtha Trod: Lumbinī and Kapilavastu*, Lumbini: Lumbinī Internation Research Institute, 2004.

Falk, Harry, *The Discovery of Lumbinī*, Lumbini: Lumbini International Research Institute (Lumbini International Research Institute, Occasional Papers, 1), 1998.

Falk, Harry, *Aśokan Sites and Artefacts – a Source Book With Bibliography*, Mainz: Philip von Zabern (Monographien zur indischen Archäologie, Kunst und Philologie, Band 18), 2006.

Falk, Harry, 'The Fate of Aśoka's Donations at Lumbinī', in Patrick Olivelle, Janice Leoshko and Himanshu Prabha Ray (eds), *Reimagining Aśoka – Memory and History*, Oxford: Oxford University Press, 2012, pp. 204–216.

Government of Nepal, Ministry of Culture, Tourism and Civil Aviation, Department of Archaeology, 'Lumbini, the Birthplace of the Lord Buddha (Nepal) (C 666rev)', *a Report on the State of Conservation of the Property*, Kathmandu, 2015.

Imam, Abu, *Sir Alexander Cunningham and the Beginnings of Indian Archaeology*, Dacca: Asiatic Society of Pakistan, 1966.

Japan Buddhist Federation, *Archaeological Research at Mayadevi Temple, Lumbini*, Tokyo: Japanese Buddhist Federation, 2001.

Japan Buddhist Federation, *Lumbini: The Archaeological Survey Report 1992–1995*, Tokyo: Japan Buddhist Federation, 2003.

Labadi, Sophia, *UNESCO, Cultural Heritage, and Outstanding Universal Value: Value-Based Analyses of the World Heritage and Intangible Cultural Heritage Conventions*, Lanham, MD and New York: AltaMira Press, 2013.

Pal, Ranajit, 'The Dawn of Religion in Afghanistan-Seistan-Gandhara and the Personal Seals of Gotama Buddha and Zoroaster', *Mithras Reader – an Academic and Religious Journal of Greek, Roman and Persian Studies*, 2010, 3: 62–83.

Phelps, Terence, *The Lumbini and Piprahwa Deception*, London: Terence Phelps Ltd., 2009.

Ringbeck, Brigitta, *Management Plans for World Heritage Sites: A Practical Guide*, Bonn: German Commission for UNESCO, 2008.

Silverman, Helaine, *Contested Cultural Heritage: Religion, Nationalism, Erasure, and Exclusion in a Global World*, New York and Dordrecht, The Netherlands: Springer, 2011.

UNESCO, *Operational Guidelines for the Implementation of the World Heritage Convention*, Paris: UNESCO World Heritage Centre, 2005, http://whc.unesco.org/archive/opguide05-en.pdf (accessed on 21 July 2016).

Weise, Kai (comp./ed.), *The Sacred Garden of Lumbini: Perceptions of the Buddha's Birthplace*, Paris: UNESCO, 2013.

9 The implementation of *Tri Hita Karana* on the world heritage of Taman Ayun and Tirta Empul temples as tourist attractions in Bali[1]

I. Wayan Ardika

Introduction

The *subak* system of Bali Province consists of rice terraces and their water temples. UNESCO inscribed this unique cultural landscape on the World Heritage List on 29 June 2012: 'The inscription recognises the value of Bali's *subaks*: farmers' organisations that collectively manage irrigation systems on rice terraces, as well as water temples. The *subak* system, which dates back to at least the 11th century, is still in practice'.[2] The nomination file emphasises that '*subaks* are not simple water-user associations managed by single communities; instead *subaks* are connected via the water temple networks into functional hierarchies that manage the landscape at different scales, from whole watersheds to individual paddies'.[3] This inscription was made possible when UNESCO introduced new concepts of living heritage such as "cultural landscape" in 1992 and "intangible heritage" in 2003. The World Heritage Committee defined cultural landscape as follows:

> cultural properties that represent the combined works of nature and man. They are illustrations of the evolution of human society and settlement over time, under the influence of the physical constraints and successive social, economic and cultural force, both external and internal.[4]

This definition led to the recognition of the non-monumental character of the heritage of cultural landscapes and to the acknowledgement of the links between cultural and biological diversity, specifically with sustainable land use. Cultural diversity guarantees sustainability because it binds universal developmental goals to plausible and specific moral visions. Biological diversity provides an enabling environment for it.[5]

Heritage specialists have argued that the categories of cultural landscape and intangible heritage reflect the priorities and concerns of nominating countries, and they have a bearing on how landscapes are preserved for posterity. For example, in the case of the *subak* system of Bali, the Hindu-Balinese philosophy of *Tri Hita Karana* means that the harmony of nature, religion and culture is considered crucial for prosperity and happiness, which parallels the notion of sustainable development,[6] a crucial aspect not only for heritage conservation, but also for society. Sustainable development of heritage sites has often been linked to tourism or more recently sustainable tourism, environmental sustainability and the sustainable development of local communities.

All the farmers who draw on a single water source whether a single dam and canal running from dam to fields, all belong to a single *subak*. To date, Bali has about 1,200 *subaks*. How is the *subak* system to be understood within the available models of the world's water systems, which have tended to focus on control of irrigation in the context of state formation?[7] Scholars have described the *subak* as a flexible multi-scale managerial system, which allows farmers to take the initiative to improve control of the ecological system. In this framework, the role of the state is limited to taxation or encouraging the expansion of the system. Archaeological investigations on the north coast of Bali and an examination of the inscriptions date rice cultivation on the island to the beginnings of the Common Era and a gradual expansion of the social structures that sustained the rice fields.[8] The earliest evidence for the presence of water temples and irrigated systems exists in southern Bali and also shows that regular meetings of *subak* heads (*pekaseh*) were held in water temples.[9]

The Balinese irrigation system is both fragile and susceptible to water theft. Its maintenance and functioning requires a high level of planning and social investment that sustains it. The Balinese philosophy of *Tri Hita Karana* (literally "three causes of prosperity") is simply translated as harmony between people and communities, people and the environment. In addition, the world of gods is the underlying principle that defines its functioning. The challenge lies in preserving the core philosophic values of the cultural landscape in order to ensure that these are not altered while Balinese culture is packaged for promoting tourism or through commodification. Sustainable development of heritage sites, especially World Heritage Sites, has often been linked to tourism or more recently sustainable tourism, environmental sustainability and the sustainable development of local communities as mentioned earlier.[10] Can the ancient Balinese philosophy become the guiding principle of sustainable cultural tourism in practice?

In a 1996 study of cultural tourism, the French anthropologist Michel Picard examined the dynamism of cultural production in touristic contexts with a focus on Bali. He argues that the emphasis on Bali as an island for the promotion of a "living heritage" or cultural tourism made the local populace self-conscious about their culture. In the post-Independence period, the Indonesian government accepted tourism as a vehicle for economic development and show-cased Bali as a model for cultural tourism. Nevertheless conscious efforts were made so as not to destroy the Balinese culture on which tourism depended and several checks and balances were put in place, so that traditional culture came to be carefully staged for tourists.[11]

Since the 1980s the Government of Indonesia has actively promoted a national culture that comprised of and was built on regional cultures. These changes and transformations have resulted in a dynamic view of culture among the Balinese as will be discussed later in this chapter. Cultural tourism is a significant revenue earner that contributes in a big way to the Balinese economy. Bali accounts for roughly 0.3% of Indonesia's land area but 37% of foreign tourist arrivals. Tourism directly employs 28% of the island's work force, and contributed nearly 30% of its GDP in 2013.[12] The issue is how to sustain cultural tourism without endangering the *subak* system and the rituals and ceremonies that form the basis of *Parhyangan* (the realm of the spirits).

There are several temples which are associated with *subaks* that have been identified as World Heritage Sites. These include Ulun Danu temple in Bangli regency, Tirta Empul temple and some archaeological sites along the Pakerisan river in Gianyar regency, Taman Ayun temple at Mengwi in Badung regency, and *Catur Angga* of Batukaru temples in Tabanan regency. These temples are considered as sources of water for several *subaks* in Gianyar, Badung and Tabanan regencies.[13] This chapter focuses on the implementation of *Tri Hita Karana* at Taman Ayun and Tirta Empul temples, which also function as tourist attractions in Bali. The issue is: How are religious (*Parhyangan*), social (*Pawongan*), and environmental (*Palemahan*) aspects of *Tri Hita Karana* implemented at Taman Ayun and Tirta Empul temples? It should be noted that the regulation of the government of Bali on tourism is also based on *Tri Hita Karana*. In other words, the Outstanding Universal Value of the Cultural Landscape of Bali Province is similar to the cultural tourism regulation of Bali Province, namely *Tri Hita Karana*.

Though the philosophy of *Tri Hita Karana* has been the guiding principle at Taman Ayun and Tirta Empul temples, touristification or commodification has occurred in relation to religious aspects. For instance, tourists are allowed to enter the temple at Taman Ayun without wearing Balinese dress, and tourists are permitted to enter the most sacred yard, or *jeroan,* at Tirta Empul although restriction for tourists is implemented in theory. Social aspects of *Tri Hita Karana* have been neglected at Taman Ayun and Tirta Empul. Information concerning the functions of shrines (*palinggih*), ceremonies and history as well as local guides are not available at both temples. Thus, the fundamental role of the Balinese temple in maintaining social harmony is not generally understood and appreciated by visitors, though the environment surrounding both temples has been managed properly. There is nevertheless scope for improvement, and toilets and rubbish bins need to be placed in the right places at Tirta Empul. The philosophy of *Tri Hita Karana* needs to be implemented properly in order to gain harmony and balance between religious, social and environment at the World Heritage Sites of Taman Ayun and Tirta Empul temples as tourist attractions in Bali.

This is qualitative research in which several informants and 60 respondents comprising of both domestic and foreign tourists were interviewed randomly. Observation and interviews were conducted, and questionnaires were distributed for data collection. Data analyses were utilised through descriptive interpretation. The result of this research indicates that three aspects of *Tri Hita Karana* – namely religious, social and natural environment – have been implemented in managing the Taman Ayun and Tirta Empul temples as tourist attractions. However, the label or branding of World Heritage Sites has not yet significantly increased the number of tourists visiting Taman Ayun and Tirta Empul temples. It should also be noted that the foreign and national tourists have not yet comprehended the outstanding values of *Tri Hita Karana.* It seems that not only World Heritage Sites in Bali need to be actively promoted as tourist attractions, but more importantly, cultural values that underwrite these structures and shrines need to be suitably disseminated.

Taman Ayun temple

Taman Ayun is the royal temple of the kingdom of Mengwi. The temple was built in 1634 by a Chinese architect, Ing Khang Choew, during the reign of I Gusti Agung Putu, the king of Mengwi, who moved his palace from Balahayu (Belayu) to Mengwi. The king of Mengwi asked

Ing Khang Cheow to build a temple in a beautiful garden. The phrase *Pura Taman Ayun* literary means a temple built in a beautiful garden.[14] The temple was abandoned by the royal family of Mengwi in 1890 during the war between the kingdoms of Mengwi and Badung. Upon the return of royal family in 1911, the temple ground was restored and returned to its original function. An earthquake in 1917 caused the collapse of several structures, and nearly 40 *desa adat* (village customary) as well as the *subak* Batan Badung contributed to the restoration to the temple.[15]

The Taman Ayun temple is located in a hilly place surrounded by canals in the eastern, southern and western parts of the site. There is a bridge connecting the entrance of the temple to the road in the southern part of the site. The ground plan of the temple of Taman Ayun looks similar to that at Angkor, representing the mount of Mahameru surrounded by the sea, as discussed in detail in Hindu mythology.

Apart from being the temple of the royal family at Mengwi, Taman Ayun also functioned as a *subak* temple. Several rites or ceremonies were normally held at the temple for *subak* purposes. For example, the holy water from the lakes was kept in the shrines at Taman Ayun temple. In addition, blessing ceremonies of the mountain gods and other fertility gods as well as the ancestors of the royal family were held at Taman Ayun temple, and then the holy water was distributed to 20 *subaks* around Mengwi. The *nangluk merana* (pest control) ceremony was also held at Taman Ayun temple. The king of Mengwi traditionally led (and still leads) the ceremony, and the prince of Mengwi also continues to perform this function. The Taman Ayun temple is also the chief water temple of the *subak* Batan Badung. The *subak* Batan Badung utilised the water from the temple's large moat, and shares responsibility for the temple with the royal family of Mengwi.

The temple of Taman Ayun is a tourist destination in Bali. This research was conducted in 2015, or three years after the temple had been inscribed on the World Heritage List. Thirty respondents, consisting of 15 foreign and 15 local or domestic tourists were identified during this research. The foreign respondents included nine women and six men. Their ages can be divided into four categories: 15–29 years old (five persons, or 33.33%); 30–44 years old (one person, or 6.67%); 45–59 years old (five persons, or 33.33%); and more than 60 years old (four persons, or 26.67%). Their occupations include businessmen, medical doctors, students and other professionals. The domestic respondents consisted of three men (20%), and 12 women (80%). The ages of domestic respondents were 15–29 years old (nine persons, or 60%); 30–44 years old (three persons, or 20%);

Table 9.1 The number of tourists who visited Taman Ayun in the last five
years

Years	Foreign tourists	Domestic tourists	Total
2014	245,940	83,751	329,691
2013	205,525	76,376	281,901
2012	111,574	62,058	173,632
2011	235,511	120,574	356,085
2010	256,442	148,278	404,720

Source: Department of Tourism, Badung Regency

45–59 years old (three persons, or 20%). There was no domestic tourist more than 60 years old. Their occupations include businessmen,
teachers and students.

Foreign respondents can be categorised as mature or senior tourists
who tend to visit religious sites or temples. All the foreign respondents
(100%) had visited Taman Ayun for the first time. It is not known
whether they will come back to visit the temple in the future. The
number of tourists who visited Taman Ayu between the years 2010 to
2014 is shown in Table 9.1.

On the basis of Table 9.1, it is evident that the number of tourists who visited Taman Ayun in the year 2012 was the lowest when
the temple was put on the World Heritage List. In the years 2013
and 2014, after Taman Ayun had been inscribed as a World Heritage
site, the number of tourists visiting the temple increased continuously.
However, the number of tourists was still lower as compared to the
year 2010, which saw 404,720 persons. This indicates that the label
or branding of World Heritage Site for Taman Ayun has not significantly increased the number of tourists visiting the temple as a tourist
attraction.

To what extent have the philosophical underpinnings of *Tri Hita
Karana* aspects at Taman Ayun been preserved and disseminated
through cultural tourism? The religious aspect (*Parhyangan*) of the
temple is strictly adhered to. Tourists are not allowed to enter the
third, inner yard of the temple. This yard is considered to be the most
sacred part of the temple, because the shrines are located here and the
ceremony is normally held in this place. However, tourists can observe
and take photographs of the shrines from outside the temple wall (see
Figure 9.1).

Figure 9.1 Main entrance of Taman Ayun temple.

Source: Courtesy of Anthony Bradbury, Wikimedia Commons

As far as the religious aspect at Taman Ayun is concerned, tourists are allowed to visit the temple without wearing traditional Balinese dress (Figure 9.2). This phenomenon is indeed a paradox, as this contradicts a regulation in Bali that states that tourists and tour guides should wear Balinese dress when they visit temples. According to the French anthropologist Michel Picard this paradox can be termed as "touristic culture". In other words, for the Balinese their culture has become on the one hand what characterises them as a specific society, and on the other hand what provides their tourist product with its distinguishing features – both an identity marker (*ciri khas*) and a trademark. Accordingly, one has grounds for inferring that the doctrine of "cultural tourism" (*pariwisata budaya*) is turning Balinese culture into what could be termed a "touristic culture" (*budaya pariwisata*) – that is, a culture characterised, according to the Balinese themselves, by a confusion between the values of culture and those of tourism – precisely what this doctrine of "cultural tourism" was initially aimed at avoiding at all costs.[16]

The social aspect (*Pawongan*) of *Tri Hita Karana* relates to the relationship between people. Since Taman Ayun emerged as a tourist

Figure 9.2 Tourists are not wearing Balinese dress while visiting Taman Ayun temple.

Source: Courtesy of the author

destination, information and services to the visitors which represent human relations become critically important. Several sign posts are available for the guidance of tourists at the temple of Taman Ayun (Figure 9.3). However, the booklet and general information concerning the temple of Taman Ayun is in the Indonesian language. There is

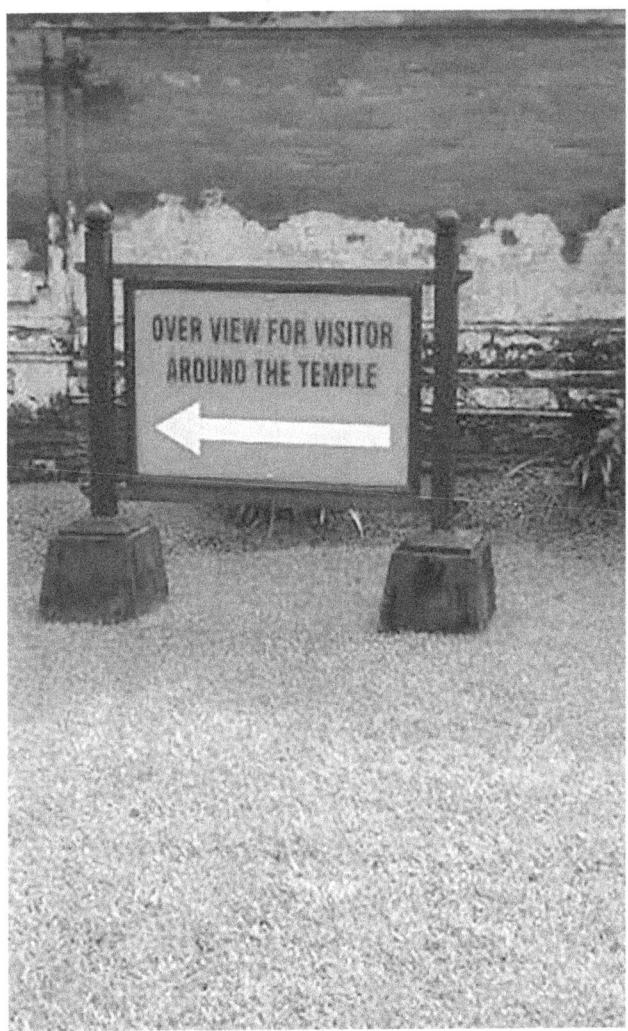

Figure 9.3 Signs at Taman Ayun temple.

Source: Courtesy of the author

no booklet or information in English or other foreign languages available at the front desk of the temple. This is inconvenient for tourists who visit the temple as casual visitors, without buying a package tour or being accompanied by a guide. In addition, there is no local guide available at the temple to provide information to the tourists.

As regards to the environmental aspect (*Palemahan*) of *Tri Hita Karana*, the gardens in the eastern and northern sides of the temple as well as the vendors in the front gate of the temple are well managed. Tourist buses are not allowed to park on the street in the front or in the south of the temple (Figure 9.4). There are several cleaners and gardeners who work at Taman Ayun to keep the lawns

Figure 9.4 Rubbish bins and gardeners at Taman Ayun temple.

Source: Courtesy of the author

green and the temple clean. The toilets and the *wantilan* (arena for cockfighting) were restored after Taman Ayun was inscribed as a World Heritage site. However, several vendors are complaining regarding a relocation programme, because buyers and tourists find it difficult to get access to their new location (Figure 9.5). This

Figure 9.5 Street management in the front of the Taman Ayun and vendors along the pedestrian walkway towards the temple.

Source: Courtesy of the author

situation must be solved by the authority as soon as possible in order to promote sustainable tourism development at Taman Ayun. The local community should benefit from tourism development which is based on cultural heritage; no one should be marginalised by the development of World Heritage management, especially the local community.

Tirta Empul temple

Tirta Empul is one of the *subak* temples along the Pakerisan river system. The springs at Tirta Empul are sources of water for *subak*s of Pulagan and Kumba in Gianyar regency. The inscription of Manukaya dated to *Saka* 882 or 960 CE described King Chandra Bhaya Singha Warmadewa as responsible for the maintenance of the springs or holy water (*tirtha*) at Air Mpul.[17] The words *tirtha di air mpul,* which were stated in the inscription, still exist today as the name of the temple: Tirta Empul.

The temple of Tirta Empul is also a tourist destination in the Gianyar regency. The number of tourists who visited the temple in the last five years, between 2010 and 2014, can be seen in Table 9.2.

Table 9.2 indicates that the number of tourists who visited the temple fluxuated in the last five years between 2010 and 2014. The number of tourists visiting Tirta Empul declined after the site was inscribed as World Heritage site.

Thirty respondents were selected randomly during the survey at Tirta Empul; 15 were foreigners and another 15 were local tourists. There are four categories of respondents in terms of their ages: 15–29 years, 30–44 years, 45–59 years, and more than 60 years. Two respondents (13.33%) of foreign tourists were in the category of 15–29 years,

Table 9.2 The number of tourists who visited Tirta Empul temple

Years	Foreign tourists	Domestic tourists	Total
2014	285,617	158,267	443,884
2013	243,459	202,043	445,502
2012	272,142	189,535	461,677
2011	188,787	177,591	366,378
2010	198,641	146,604	345,245

Source: Department of Tourism, Gianyar Regency

while six respondents (40%) were 30–44 years old. Two respondents (13.33%) were in the category of 45–59 years old, and five respondents (33.34%) in the category more than 60 years old. In terms of gender the foreign respondents consisted of four men and 11 women. Their occupations include entrepreneurs, teachers, medical doctors, student, pensioners and shopkeepers. Seventy-three percent of foreign tourists bought package tours to the temple or destination, 6.67% knew the site through a friend, 6.67% through books and 13.33% through the internet.

The domestic (local and national) respondents can be classified as follows: 15–29 years, three respondents (20%); 30–44 years, six respondents (40%); and 45–59 years, six respondents (40%). No domestic respondents were older than 60 years. Most foreign and domestic respondents were not aware that the temple had been inscribed as a World Heritage Site which is based on the philosophy of *Tri Hita Karana*.

Unlike Taman Ayun, every tourist who visits Tirta Empul temple must wear Balinese dress. Sarongs and scarves are provided for tourists at the entrance to the temple (see Figure 9.6).

Tourists who visit Tirta Empul must dress in traditional clothes in order to keep the sanctity of the temple. However, tourists at Tirta Empul are allowed to enter the third yard or the most sacred part of the temple (see Figure 9.7). This phenomenon is totally different as compared to the situation at Taman Ayun, where tourists are forbidden to enter the third yard of the temple itself. At Tirta Empul, only certain areas of the inner yard or the third yard of the temple are restricted for the tourists (see Figure 9.8). As expressed in Picard's words:

> This movement takes place by reconciling simultaneously tourism to culture and culture to tourism. On the one hand, the very fact of qualifying tourism as "cultural" bestows it with the attributes of culture, thereby exorcising the threat of destruction that it carries and legitimising its penetration of Bali. But this is not enough: while it is stressed that tourism must become "cultural" in order to be acceptable to the Balinese, it is just as necessary that their culture be marketable as a tourist product.[18]

This implies that culture must bear the attributes of tourism.

The social aspect (*Pawongan*) of *Tri Hita Karana* at the temple of Tirta Empul was represented by providing services and information to

Figure 9.6 Sarongs and scarves are provided at the entrance of Tirta Empul temple.

Source: Courtesy of the author

the tourists. As already noted, sarongs and scarves were provided to the tourists when they visit the temple. Signages and information notices are also provided to the tourists. In addition, tourists who are willing to purify themselves at the holy springs of the temple are welcome (see

Figure 9.7 Tourists taking photographs of themselves at the inner yard of Tirta Empul temple.

Source: Courtesy of the author

Figure 9.9). There is a myth that the holy springs were created by the god Indra to prevent his followers from poison which was produced by King Mayadanawa. Not only the local people but also foreign tourists take part in *melukat* (a purification ceremony). The tourists bring offerings and wear Balinese costumes for this ceremony. This

Figure 9.8 Notice forbidding entry to the inner yard of Tirta Empul temple.
Source: Courtesy of the author

behaviour represent "cultural tourism", namely tourists who respect and practice local traditions.

It should be noted that tourists complain about the fees and condition of toilets and the places for changing clothes after they have had the purification ceremony. The toilet fees seem too commercial for the tourists and the condition of the toilets is below international standards.

Figure 9.9 Tourists make queues and practice *melukat* (purification ceremony) at Tirta Empul.

Source: Courtesy of the author

The environment aspect (*Palemahan*) of *Tri Hita Karana* at the temple of Tirta Empul has been managed properly. Pedestrian spaces were created that made it easier for visitors to observe the temple. The *wantilan* pavilion was restored and the fish pond was maintained as an additional tourist attraction (see Figure 9.10).

Figure 9.10 Pedestrians around the temple and the fish pond at Tirta Empul.
Source: Courtesy of the author

Discussion

Taman Ayun and Tirta Empul have been inscribed as World Heritage Sites. Cultural heritage is contested and negotiated in the interplay of local, national and global spaces.[19] The philosophy of *Tri Hita Karana* as the Outstanding Universal Values or local wisdom of the Balinese people should be understood and experienced by both national and foreign tourists. As already noted, a majority of national and foreign tourists have not yet understood the Outstanding Universal Values of *Tri Hita Karana*. It is evident that the tourism authority of the government of Bali and the managements of both Taman Ayun and Tirta Empul should promote the philosophy of *Tri Hita Karan* to tourists and travel agents in order to enrich the cultural experience of the tourists visiting those temples.

As mentioned earlier, the number of tourists visiting Taman Ayun and Tirta Empul after the temples were inscribed as World Heritage Sites has not increased significantly (see Tables 9.1 and 9.2). A similar case also occurred at Borobudur. The number of foreign tourists who visited Borobudur in 2014 was 250,000 persons. The Minister of Tourism of Indonesian government stated that Borobudur should aim for 2 million tourists. The Angkor Wat temple in Cambodia was visited by 2.1 million foreign tourists in 2015.[20] In comparison to the old city of Lijang in China the situation is quite different. According to the Tourism Board of Lijiang City, 1.06 million tourists visited the Old Town in 1996, one year before the World Heritage designation, and this number rose to 4.33 million in 2007. In 2012 the number of visitors amounted to 16 million. Most of them are Chinese domestic tourists. In terms of tourism development, the Old Town seems quite successful, and the effects of World Heritage are considerable.[21] Many writers criticise the effects of World Heritage designation from the viewpoint of Lijiang's local community. They write, "Lijiang does not need World Heritage". Through the increase in number of tourists, the river water, which had supported the lives of the local people, is now polluted. Traditional houses in the Old Town have been transformed into souvenir shops, restaurants or guesthouses. The Naxi people who live in the Old Town have moved out and Han Chinese have moved in to run the shops. As of 2007, Naxi residents ran only 98 of 379 guesthouses in the Old Town, while 281 were run by non-Naxi, mostly Han Chinese. In other words, the Old Town of Lijiang today exists for the tourists rather than the Naxi.[22]

It is interesting to note that the government of Bali Province has developed a Cultural Tourism policy which promotes the cultural identity of a local population as a tourist attraction.[23] The impacts of tourism industry on Bali were understood to represent a threat of cultural pollution. To prevent such an adverse outcome, the Balinese authorities devised a policy of Cultural Tourism which was intended to develop tourism without debasing Balinese culture, by using culture to attract tourists while fostering culture through the revenue generated by tourism.[24]

The use of Taman Ayun and Tirta Empul temples as tourist attractions in Bali has had some impacts on the religious, social and environment aspects. There are restrictions or limitations on access for tourists to Taman Ayun and Tirta Empul in order to maintain the sanctity of the temples. However, allowing tourists without wearing Balinese dress to enter the Taman Ayun temple and tourists to freely visit the inner yard of the Tirta Empul may be identified as touristification processes. In other words, the sacredness of the temples of Taman Ayun and Tirta Empul appears to be declining due to the tourism industry. This situation should be addressed by negotiating and adopting local values by tourists in order to maintain the sacredness or religious aspect the temples.

Shinji Yamashita states that the central issue of world cultural heritage has highlighted the interplay of local, national and global interests. The question is, who will make use of heritage site, for whom, and for what purpose? Priority should be given to the local agents in cultural resources management. Cultural resources should be utilised primarily for local benefit. National and international agents should only be collaborators in the exploitation of cultural resources, not the main beneficiaries. It is clear that World Heritage designations should be used primarily for the benefit of the local communities where World Heritage Sites are located, contributing to their happiness.[25]

In relation to the tourism industry, tourism was accused of corrupting Balinese culture, desacralising temples and profaning religious ceremonies, monetising social relations and weakening communal ties, or relaxing moral standards and promoting the rise of mercantile attitudes. The issue was about what could be presented and marketed to the tourists. In this case, culture was being designed as a product.[26] As World Heritage Sites, Taman Ayun and Tirta Empul are no longer the exclusive property of the Balinese alone. Both temples become main attractions for tourists, as brand images of tourist destinations. According to Picard, culture or the Balinese temples had become capital, which they could exploit for a profit.[27] It seems that one could speak of "touristic culture" once the Balinese came to confuse these two uses of their culture, when that by which the tourists identified them became synonymous with that through which they identified themselves – that is, when the imperatives of the touristic promotion of their culture informed the considerations of their motivation to preserve it, to the extent that the Balinese ended up mistaking the brand image of their tourist product for the maker of their cultural identity

Indeed as soon as a society offers itself for sale on a market, as soon as it attempts to enhance its appeal to the eyes of foreign visitors, then the very consciousness that society has of itself is affected. In this respect, local people are not passive objects of the tourist gaze, but active subjects who construct representations of their culture to attract tourists.[28]

Conclusion

The designation of World Heritage Sites of Taman Ayun and Tirta Empul has not yet significantly increased the number of tourists visiting the temples. The outstanding values of *Tri Hita Karana* – namely harmony and balance between religious (*Parhyangan*), social (*Pawongan*), and environmental (*Palemahan*) aspects – should be promoted and maintained for the benefit of local people.

Contestation and negotiation at local, national and international levels should be conducted at Taman Ayun and Tirta Empul as World Heritage Sites. Taman Ayun and Tirta Empul represent the cultural identity of the Balinese and should be preserved and well as serve as capital for the tourists. In this case, World Heritage Sites should prove beneficial for the local people and the tourists as well.

Notes

1 An earlier version of this paper was presented at the International Tourism Conference Promoting Cultural & Heritage Tourism, Udayana University, 1–3 September 2016.
2 Albert M. Salamanca, Agus Nugroho, Maria Osbeck, Sukaina Bharwani and Nina Dwisasanti, *Managing a Living Cultural Landscape: Bali's Subaks and the UNESCO World Heritage Site*, Bangkok: Stockholm Environment Institute, Asia, 2015, p. 1. The official name of the site is the Cultural Landscape of Bali Province: the Subak System as a Manifestation of the Tri Hita Karana Philosophy. For a detailed description and background, see 'Cultural Landscape of Bali Province: the Subak System as a Manifestation of the Tri Hita Karana Philosophy', https://whc.unesco.org/en/list/1194 (accessed on 28 June 2018).
3 Ministry of Culture and Tourism, Republic of Indonesia and the Government of Bali Province, 2011, Cultural Landscape of Bali Province, Nomination for inscription on the UNESCO World Heritage List. 'Cultural Landscape of Bali Province: the Subak System as a Manifestation of the Tri Hita Karana Philosophy', https://whc.unesco.org/en/list/1194 (accessed on 28 June 2018).
4 Shinji Yamashita, 'The Balinese *Subak* as World Cultural Heritage: In the Context of Tourism', in I. Nyoman Darma Putra and Siobhan Campbell (eds), *Recent Developments in Bali Tourism: Culture, Heritage, and landscape in an Open Fortress*, Buku Arti: Denpasar, 2015, p. 120.
5 'Cultural Diversity and Biodiversity for Sustainable Development', http://unesdoc.unesco.org/images/0013/001322/132262e.pdf (accessed on 28 June 2018).
6 Keiko Miura and I. Made, 'The World Heritage Nomination of Balinese Cultural Landscape: Local Struggles and Expectations', in Victor T. King (ed.), *UNESCO in Southeast Asia: World Heritage Sites in Comparative Perspective*, Copenhagen: Nordic Institute of Asian Studies Press, 2016, pp. 274–290.
7 J. Stephen Lansing, Murray P. Cox, Sean S. Downey, Marco A. Janssen and John W. Schoenfelder, 'A Robust Budding Model of Balinese Water Temple Networks', *World Archaeology*, March 2009, 41(1), The Archaeology of Water: 112–133.
8 Brigitta Hauser-Schäublin and I. Wayan Ardika (eds), *Burials, Texts and Rituals: Ethnoarchaeological Investigations in North Bali, Indonesia*, Göttingen: Universitätsverlag Göttingen, 2008.
9 Lansing et al., 'A Robust Budding Model of Balinese Water Temple Networks', pp. 114–116.
10 Miura and Made, 'The World Heritage Nomination of Balinese Cultural Landscape', pp. 274–290.
11 Michel Picard, *Cultural Tourism and Touristic Culture*, Singapore: Archipelago Press, 1996, pp. 120–121, www.espacestemps.net/en/articles/bali-the-discourse-of-cultural-tourism/ (accessed on 7 July 2017).
12 Salamanca et al., *Managing a Living Cultural Landscape*, p. 4.
13 Stephen J. Lansing and Julia N. Watson, 'Guide to Bali's UNESCO World Heritage: "Tri Hita Karana: Cultural Landscape of Subak and Water Temple"', Final draft 2012, publication pending, 2012.

14 I. Wayan Ardika et al., *Implementasi Tri Hita Karana dalam Pengelolaan Warisan Budaya Dunia sebagai Daya Tarik Pariwisata di Bali*, Denpasar: Laporan Akhir Hibah Grup Riset, Universitas Udayana, 2015, p. 11.
15 Lansing and Watson, 'Guide to Bali's UNESCO World Heritage', pp. 86–87.
16 Michel Picard, 'Cultural Tourism', in *Bali: Cultural Performances as Tourist Attraction*, p. 74, https://cip.cornell.edu/DPubS?service=Repository&versio n=1.0&verb=Disseminate&view=body&content-type=pdf_1&handle=seap. indo/1107012381# (accessed on 7 July 2017).
17 Ardika et al., *Implementasi Tri Hita Karana dalam Pengelolaan Warisan Budaya Dunia sebagai Daya Tarik Pariwisata di Bali*, p. 16.
18 Picard, *Cultural Tourism and Touristic Culture*.
19 Yamashita, 'The Balinese *Subak* as World Cultural Heritage', p. 139.
20 Sor Chandara, 'Angkor Visitors Number Flat in 2015', ww.phnompenhpost. com/business/angor-visitor-numbers-flat-2015 (accessed on 28 June 2018).
21 Yamashita, 'The Balinese *Subak* as World Cultural Heritage', p. 130.
22 Yamashita, 'The Balinese *Subak* as World Cultural Heritage', pp. 130–131.
23 Michel Picard, 'Balinese Identity as Tourist Attraction: From "Cultural Tourism" to "Ajeg Bali", in I. Nyoman Darma Putra and Siobhan Campbell (eds.), *Recent Developments in Bali Tourism: Culture, Heritage, and Landscape in an Open Fortress*, Denpasar: Buku Arti, 2015, pp. 39–68, 45.
24 Picard, *Balinese Identity as Tourist Attraction*, p. 47.
25 Yamashita, *The Balinese Subak as World Cultural Heritage*, pp. 134–140.
26 Picard, *Balinese Identity as Tourist Attraction*, p. 50.
27 Picard, *Balinese Identity as Tourist Attraction*, p. 51.
28 Picard, *Balinese Identity as Tourist Attraction*, p. 52.

Bibliography

Ardika, I. Wayan, I. Nyoman Dhana, and dan I. Ketut Setiawan, *Implementasi Tri Hita Karana dalam Pengelolaan Warisan Budaya Dunia sebagai Daya Tarik Pariwisata di Bali*, Denpasar: Laporan Akhir Hibah Grup Riset, Universitas Udayana, 2015.

Chheang, Vannarith, 'Angkor Heritage Tourism and Tourist Perceptions', *Tourismos: An International Multidisciplinary Journal of Tourism*, 2011, 6(2): 213–240.

Cooper, Chris, John Fletcher, Alan Fyall, David Gilbert, and Stephen Vanhill, *Tourism Principles and Practice*, 3rd edition. Edinburgh Gate: Pearson Education Limited, 2005.

Grader, G. J., 'The State Temples of Mengwi', in W. F. Dalam Wertheim (ed.), *Bali Studies in Life, Thought, and Ritual*, The Hague, The Netherlands and Bandung: W. Van Hoeve Ltd., 1960, pp. 155–186.

Hardesty, Donald L., and Barbara J. Little, *Assessing Site Significance*, New York: AltaMira Press, 2009.

Hauser-Schäublin, Brigitta, and I. Wayan Ardika (eds), *Burials, Texts and Rituals: Ethnoarchaeological Investigations in North Bali, Indonesia*, Göttingen: Universitätsverlag Göttingen, 2008.

Hitchcock, M., Victor T. King, and Michael Parnwell (eds), *Heritage Tourism in Southeast Asia*, Singapore: Nias Press, 2010.

Lansing, Stephen J., Murray P. Cox, Sean S. Downey, Marco A. Janssen, and John W. Schoenfelder, 'A Robust Budding Model of Balinese Water Temple Networks', *World Archaeology*, March 2009, 41(1), *The Archaeology of Water*: 112–133.

Lansing, Stephen J., and Julia N. Watson, 'Guide to Bali's UNESCO World Heritage: "Tri Hita Karana: Cultural Landscape of *Subak* and Water Temple"', Final draft 2012, publication pending, 2012.

Lipe, William, 'Value and Meaning in Cultural Resource', in H. Dalam Cleere (ed.), *Approaches to the Archaeological Heritage*, Cambridge: Cambridge University Press, 1984, pp. 1–11.

Madiasworo, Taufan, Gunawan Tjahjono, Budhy Tjahjati, and Subur Budhisantoso, 'Sustainable Heritage Area Management Model Study on Environmental Wisdom in Taman Ayun Area, Badung Regency, Bali Province', *Australian Journal of Basic and Applied Sciences*, 2014, 8(10): 219–225.

Miura, Keiko, and I. Made, 'The World Heritage Nomination of Balinese Cultural Landscape: Local Struggles and Expectations', in Victor T. King (ed.), *UNESCO in Southeast Asia: World Heritage Sites in Comparative Perspective*, Copenhagen: Nordic Institute of Asian Studies Press, 2016, pp. 274–290.

Nordholt, Henk Schulte, 'Representing Traditional Bali: Colonial Legacies and Current Problems', in I. Nyoman Darma Putra and Siobhan Campbell (eds), *Recent Developments in Bali Tourism: Culture, Heritage, and Landscape in an Open Fortress*, Denpasar: Buku Arti, 2015, pp. 1–10.

Pemerintah Provinsi Bali, *Peraturan Daerah Provinsi Bali Nomor 2 Tahun 2012: Tentang Kepariwisataan Budaya Bali*, Denpasar: Pemerintah Provinsi Bali, 2012.

Picard, Michel, *Bali: Pariwisata Budaya dan Budaya Pariwisata*, Jakarta: Kepustakaan Populer Gramedia dan Ecole francaise d'Extreme-Orient, 2006.

Picard, Michel, 'Balinese Identity as Tourist Attraction: From "Cultural Tourism" to "Ajeg Bali"', in I. Nyoman Darma Putra and Siobhan Campbell (eds), *Recent Developments in Bali Tourism: Culture, Heritage, and Landscape in an Open Fortress*, Denpasar: Buku Arti, 2015, pp. 39–68.

Salamanca, Albert M., Agus Nugroho, Maria Osbeck, Sukaina Bharwani, and Nina Dwisasanti, *Managing a Living Cultural Landscape: Bali's Subaks and the UNESCO World Heritage Site*, Bangkok: Stockholm Environment Institute – Asia, 2015.

Setiawan, I. Ketut, 'Komodifikasi Pusaka Budaya Pura Tirta Empul dalam konteks Pariwisata Global', *Disertasi*, Denpasar: Program Pascasarjana Universitas Udayana, 2011.

Surata, Sang Putu Kaler, *Lanskap Budaya Subak. Belajar dari masa lalu untuk membangun masa depan*, Denpasar: Universitas Mahasaraswati Press, 2013.

Yamashita, Shinji, 'The Balinese *Subak* as World Cultural Heritage: In the Context of Tourism', in I. Nyoman Darma Putra and Siobhan Campbell (eds), *Recent Developments in Bali Tourism: Culture, Heritage, and Landscape in an Open Fortress*, Denpasar: Buku Arti, 2015, pp. 116–144.

Internet

Sor Chandara, 'Angkor Visitors Number Flat in 2015', www.phnompenhpost. com/business/angor-visitor-numbers-flat-2015

10 Transnational heritage

Building bridges for the future

Himanshu Prabha Ray

Transnational heritage ties in with UNESCO's agenda of shifting attention from national histories to globalisation. It also offers an opportunity to underscore cultural diversity both in a local context as well as in the global arena. As co-signatories of the UNESCO Convention, nation states have spearheaded issues of cultural heritage, but transnational nominations provide the prospect to move beyond state-to-state relations and to use cultural heritage resources both for economic development and more importantly for understanding environmental change, as well as enhancing international collaboration in the cultural sector. Transnational World Heritage nominations are an innovative approach of recognising many cultural heritage sites located in different countries under the umbrella of one property containing Outstanding Universal Value. This chapter analyses the role of UNESCO's World Heritage Centre[1] and its Advisory Bodies such as the International Council on Monuments and Sites (ICOMOS) and International Centre for the Study of the Preservation and Restoration of Cultural Property (ICCROM) in assisting and promoting Transnational World Heritage Nominations. Recent inscriptions have however raised several issues: Is transnational nomination about inscription of as many sites as possible or is it about protection of universal values that would help focus a divided world on a shared heritage? It also raises the responsibility of maintaining equity in the inscription of transnational nominations among different partners. Are UNESCO's frameworks (Outstanding Universal Value, or OUV) designed for promoting a worldview weighted towards powerful nation states or can these be negotiated to result in meaningful dialogue for appreciation of universal cultural values?

Transnational heritage also draws into its ambit multilateral development banks that could be tapped for financial support, such as the World Bank and regional organisations such as the Asian

Development Bank.[2] It is this opportunity to build networks outside the domain of the States Parties that makes transnational nominations critical to UNESCO's agenda for establishing cultural diversity. Cultural diversity has been understood from a gender perspective. Indeed, the mainstreaming of women has been at the heart of the agenda of the United Nations and its specialised agencies, particularly UNESCO, for the past 30 years, though this is an issue seldom discussed in the South Asian context. In addition, cultural diversity has been understood in terms of the representation and participation of local communities.[3]

However, compared to single-site nominations, serial transnational nominations face three common challenges: data handling is more complex; policies and heritage management might differ; and more stakeholders are involved. Categories of heritage within the framework of the World Heritage Convention have been broadened over the decades to include cultural landscapes, industrial remains, and heritage routes, which are all now valued as part of our cultural heritage. The World Heritage Committee has also reflected upon such subjects as the need for community involvement, social benefits, heritage as part of sustainable development and the engagement of young people in the World Heritage process. The UNESCO Universal Declaration of Cultural Diversity, adopted in November 2001, states

> Culture takes diverse forms across time and space. This diversity is embodied in the uniqueness and plurality of the identities of the groups and societies making up humankind. As a source of exchange, innovation and creativity, cultural diversity is as necessary for humankind as biodiversity is for nature.[4]

How does this global vision of a plural cultural landscape as envisaged by UNESCO relate to international ambitions of presenting world or global history, especially when research on ethnicity and nationalism has argued for the importance of heritage in community and identity making?

Heritage is seen as a powerful diplomatic tool worldwide that is considered above the coercive sphere of politics, economics or military aspirations.

> While UNESCO's World Heritage programme has a global vision and remit, its work however reinforces the interests of nation states and is closely tied in to economic benefits, national identity and prestige. National delegations to UNESCO are led not by experts,

but by career diplomats, thus further corroborating the growing importance of World Heritage within nationalist agendas.[5]

Is it possible to incorporate national goals with global and universal values? What role do bodies such as ICOMOS and ICCROM play in promoting transnational agendas? This chapter addresses these issues by examining the history and archaeology of the first transnational World Heritage Site in India. The chapter is in three sections. In the first part, I discuss the first example of transnational heritage from India; while in the second, the attempt is to place another of India's possible nominations on maritime heritage within a larger framework of China's push towards a Maritime Silk Routes/Road nomination. The final section discusses a shipwreck site from the subcontinent with close links to a political centre and a monastic complex and its implications for transnational heritage.[6]

Locating transnational heritage in India

The first transnational site in India was inscribed on the World Heritage List in 2016 in Turkey and titled Architectural Work of Le Corbusier (1887–1965), an Outstanding Contribution to the Modern Movement. It comprises 17 sites spread over seven countries: Argentina, Belgium, France, Germany, India, Japan and Switzerland. Of these, ten properties are in France, two in Switzerland and one each in the other countries. Clearly this does not make for an equitable distribution of sites. The sites were chosen from the work of Le Corbusier built over a period of a half century and are a testimonial to the invention of a new architectural language that made a break with the past. The structures were chosen as they represent 'the challenges of inventing new architectural techniques to respond to the needs of society'.[7] In the context of India, the Chandigarh Capitol complex, comprising four edifices (the High Court, the Legislative Assembly, the Secretariat and the Museum of Knowledge) was selected as representative of Le Corbusier's European modernism adapted to suit Indian geographical location.[8] While this nomination submitted by the Chandigarh Department of Tourism accepts that Chandigarh is one of 14 other contemporaneous new Indian towns, there is no mention of either the cultural or historical aspect of Chandigarh. Is architecture only to be defined in terms of European modernism or does it draw from a cultural context?

What this inscription misses completely is the social history of urban planning in post-Independence India. It is people and communities who make a city, rather than urban planning and architecture. In March 1948, the Government of Punjab, in consultation with the Government of India, approved the area at the foothills of the Shivaliks as the site for the new capital, which the French architect Le Corbusier was entrusted to design. In his detailed study of the three post-Independence cities of India – i.e. Bhubaneshwar, Chandigarh (1951–1965) and Gandhinagar – Ravi Kalia suggests that:

> as a result of political changes stemming from independence and partition, India was forced to build new state capitals and add extensions to existing cities to provide homes to refugees, house state governments, and deal with urban congestion. . . . The development of Chandigarh, Bhubaneswar and Gandhinagar, between 1949 and 1982, represents a fascinating study of practical politics, personal ambitions of politicians and Western planners, and the high ideals of Prime Minister Jawaharlal Nehru. . . . The story of Chandigarh, Bhubaneswar and Gandhinagar is not one of success or failure or even of comparative satisfaction with the quality of life in a new city. It is, rather, a chronicle of a period during which India made a bold attempt to make a break with her past within the confines of a socio-urban experiment that included, along with an innovative master plan, modernist buildings, new land-use patterns, provisions for education, recreation, medical and social services, the careful and deliberate inclusion of ideas that had their origin in a culture far removed from her own.[9]

The three cities did not develop from the same contingencies in independent India. After the partition of Punjab on 15 August 1947, not only did India lose Lahore – the erstwhile capital to Pakistan – but the State itself was left deeply scarred with the psychological trauma of the more than 4.9 million refugees who came from West Pakistan into India.[10] The issue of housing the refugees was urgent and immediate; it was felt that no existing city in the Punjab could handle the influx. Three considerations were significant in the planning of the new city: strategic and military security against hostile Pakistan, adequate space for expansion and the potential to recover from the cultural loss of Lahore.[11] Perhaps the crucial question of cultural loss remains unanswered by the planners and authorities of the new city and those that proposed it for transnational nomination. At present the city of Chandigarh is bitterly divided along linguistic lines and attempts at maintaining the rural–urban divide are long lost.

The case of Chandigarh also highlights the fact that the planning of the cities is also symptomatic of the multiple agencies involved and their often contradictory agendas. Jawaharlal Nehru wrote to the then Premier of East Punjab on December 7, 1949:

> there is too great a tendency for our people to rush up to England and America for advice. The average American or English townplanner will probably not know the social background of India. He will therefore be inclined to plan something which might suit England or America, but not so much India.[12]

Pointing to New Delhi as an example, he concluded, 'This is attractive in a way, but most inconvenient and most un-Indian'. He strongly recommended the choice of the German architect Otto Koenigsberger (1909–1999), who was then already in India and an employee of the Government of India, or the American planner and architect Albert Mayer (1897–1981), who was employed in several projects in India such as the Master Plan for Greater Bombay.[13]

Mayer was appointed to prepare an urban plan for Chandigarh in December 1949, and he proposed a fan-shaped plan. Without deciding on Mayer's plan, a second architectural adviser – Le Corbusier – was appointed in December 1950. Neither Mayer nor Le Corbusier agreed to spend any length of time in India, with the latter suggesting that he would make two visits to India annually, each of one-month duration! When Mayer enquired about his project, he was informed that a group of new architects had been engaged from France and England – viz. Maxwell Fry (1899–1987), Jane Drew (1911–1996), and Le Corbusier and his cousin Pierre Jeanneret (1896–1967) – but as a reassurance was told that all architects hired endorsed the principles of his plan. This however was not to be the case. Le Corbusier introduced his own ideas, retaining nevertheless all the distinctive features of the Mayer Plan. 'With Mayer reduced to a simple footnote in the history of Chandigarh, it would be Le Corbusier who would be popularly remembered as the creator of the city'.[14]

For both Fry and Le Corbusier, industrialisation was the culprit that had irrevocably altered the character of contemporary cities. New forms in city living had to be developed to mitigate the pernicious effects of industrialisation, new forms that would herald the beginning of the Second Machine Age.[15] The Capitol Complex at Chandigarh is seen as Le Corbusier's most spectacular work. The monument of the Open Hand symbolised the beginning of the Second Machine Age: 'Open to receive the newly created wealth, open to distribute it to its people and to others'.[16] The main material of construction was raw concrete, which Le Corbusier had discovered in 1908.

Was this break with the past successful? What place did heritage and remnants of the past have in these new cities? How does the planned city designed by a French modernist fulfil the role of a regional capital commissioned by a new nation state in search of a post-colonial identity? Unfortunately, these issues have found no space in this transnational inscription. Why should the name of the modernist city – Chandigarh – derive from the temple of goddess Chandi located in the vicinity of the site selected for the city?

One of the buildings that the French architect planned for the new city was that of the Government Museum and Art Gallery. Situated close to the city centre, the museum has a sprawling and extensive campus at one side of which is located the Government College of Art. The museum was inaugurated on 6 May 1968 under the initiative and active support of Dr. M. S. Randhawa (1909–1986), renowned connoisseur and patron of art, and the then Chief Commissioner of Chandigarh. The Punjab Legislature passed the Punjab Ancient and Historical Monuments and Archaeological Sites and Remains Act, 1964 and around the same time the state established an archaeological cell under the Director Archives and Curator Museums. It was this newly established cell that started excavations at Sanghol under the charge of R. S. Bisht on 20 December 1968 and continued until 1974. Thus, the development of the Government Museum, Chandigarh was coterminous with the ongoing archaeological excavations at Sanghol. The museum boasts a rich collection of antiquities proudly displaying the long and rich heritage of Chandigarh, most of which was marginalised in the euphoria of establishing a modern city in the post-Independence period. But has this neglect been corrected through the recent transnational nomination?

Fifty kilometres from Chandigarh is the Harappan site of Ropar, the first Harappan site excavated in independent India. Harappan remains were found while digging for the new city in the 1950s and 1960s and are now displayed in the Government Museum and Art Gallery in the city. In December 1969, while digging the foundations of a building in the shopping area of Sector 17C, in the centre of Chandigarh, the remains of a Harappan cemetery were unearthed, as also a settlement located about 100 metres to the east-northeast of the cemetery.[17] The riches from the past around Chandigarh date to the pre-historic period as fossils have been found at Masoul village near Mohali by a joint Indo-French team. These were displayed in the government museum at Chandigarh to show to the visiting French president in May 2016.

The Government Museum at Chandigarh is also home to another large and important collection. In 1949, of the 962 Gandharan pieces in the Lahore Museum in Pakistan, 627 pieces were transferred to East Punjab in India along with 92 ancient sculptures from other periods and 447 miniature paintings from the Punjab Hills.[18] How do these museum collections address themselves to identity and cultural traditions of the present state of Punjab? Fifty kilometres west of Chandigarh is the Buddhist archaeological site of Sanghol, where excavations have yielded evidence of settlement from the second millennium BCE to the 12th century CE. Brick-built stupas were found at the site. In 1985, the excavators discovered beautifully sculpted stone railings of a stupa buried neatly in its vicinity (Figures 10.1 and 10.2). This discovery brought Sanghol back into the limelight and also drew the attention of institutions of the Government of India, such as the National Museum and the Archaeological Survey of India. An exhibition of selected pieces of railing pillars from Sanghol was arranged in the National Museum, New Delhi, which was inaugurated by the president of India. A catalogue of sculptures from the site of Sanghol was published by the Department of Cultural Affairs, Punjab, jointly with the National Museum.[19] Twelve

Figure 10.1 The brick stupa excavated from the site of Sanghol near Chandigarh.
Source: Courtesy of the author

Figure 10.2 Railing pillars from the Buddhist site of Sanghol, now in the site museum.

Source: Courtesy of the author

of the railing pillars from Sanghol are displayed at the Chandigarh Museum. Clearly Chandigarh had a much longer engagement with the past than that presented in this transnational nomination. The nomination dossier refers to Chandigarh as a 'unique symbol of the progressive aspirations of the new republic and the ideology of its struggle for independence'; 'first post-colonial city in India to provide a generous cultural and social infrastructure'. However, these attributes have very little to show on the ground and the reality of this transnational inscription does not support the statements made in the dossier. It would be best to conclude this section with a quote from V. S. Naipaul:

Le Corbusier's unrendered concrete towers, after 27 years of Punjab sun and monsoon and sub-Himalayan winter, looked stained and diseased, and showed now as quite plain structures, with an applied flashiness: megalomaniac architecture: people reduced to units, individuality reserved only to the architect, imposing his ideas of colour in an inflated Miroesque mural on one building and imposing an iconography of his own with a giant hand set in a vast flat area of concrete paving, which would have been unbearable in winter and summer and the monsoon. India had encouraged yet another outsider to build a monument to himself.[20]

Another example of "heritage tourism" aimed at Europeans is evident in the case of Tranquebar or Tharangambadi, a small fishing town on the coast of Tamil Nadu, which was a Danish trading colony from 1620 to 1845, when it was sold to the British (Figure 10.3). The first Protestant church was founded in Tranquebar in 1707. In 1978 the Department of Archaeology, Government of Tamil Nadu took over the control of Fort Dansborg and in 1980 Tranquebar was declared a heritage town by the Government of Tamil Nadu. Subsequently in 2008, the Union Ministry of Tourism decided to develop it as a major tourist destination, further supported by the Tranquebar Initiative of the National Museum of Denmark, a large cross-disciplinary research initiative. This has generated large-scale conservation and restoration efforts aimed at preserving the monuments of the town's colonial past, such as the former Danish governor's residence.[21] Unfortunately these attempts at restoration of the "colonial" past seem to be aimed at European tourists who flood the town and are major revenue earners for the local populace. Amongst the Danish non-governmental organisations that take part in restoration projects in Tranquebar the interest is largely because the town has a special significance as a trading post or overseas colony of the Danes.

Figure 10.3 Danish fort at Tharangambadi.
Source: Courtesy of the author

What is forgotten in the process is the pre-European settlement at the town and the complex relationship between the trading post and the kings of Tanjore to whom the Danes paid tribute. An early 14th century Masilamani Nathar temple to Shiva is known from Tranquebar.[22] During the 17th century the Tanjore court was a centre for performing arts, and it patronised the *devadasis* (female dancers) serving at temples not only in Tanjore, but also in larger towns such as Tiruvarur, 40 kilometres southwest of Tranquebar and at Tranquebar as well. Thus, Tranquebar was a part of the cultural network of Tanjore and its ruling elite.

This larger maritime landscape was by no means the result of European initiative, as evident from the history of the town of Nagapattinam, 35 km south of Tranquebar. In the 11th and 12th centuries, the king of Srivijaya and Chinese patrons made donations to the Buddhist temple at Nagapattinam. Nor should this be seen as an isolated event, but as a continuation of a long cultural process across the route connecting the southeast coast of India to Sumatra, Vietnam and China.[23] Markers of this longevity include early coastal temples and their precursors, the Iron Age burials. Rather than focus on the inscriptions and the structure of the kingdom, it is important to trace the archaeological beginnings of settlement in the area, relationship with inland Megalithic groups, the inter-island networks of the southeast coast of Sumatra, and finally, the wider linkages of the Sriwijayan kingdom with South Asia. This is a theme for subsequent research, but here I would like to raise the issue of community participation and regional heritage in projects perceived at targeting European tourists.

How is the resurrection of the Danish heritage perceived by the locals – as the location of cross-cultural encounters or as elite activity that marginalises the fishing communities of the village? Does the local community distinguish between the Danes and the British who followed them? Which heritage is being preserved and for whom? The answer to these questions is evident from recent buildings that have come up in the area. Tharangambadi was struck by the tsunami in December 2004 and as a result, the houses of the large fishing community were moved inland. The space vacated by the fishing groups on the beach which they had earlier used for drying fish has now been occupied by hotels catering to "heritage tourism" by Europeans.[24] Between 2004 and 2008 three heritage hotels were built in Tharangambadi and the Indian National Trust for Art and Cultural Heritage (INTACH) has preserved several buildings of the colonial period on Goldsmith Street in collaboration with a private funding agency (Figure 10.4):[25]

> Thus five Tamil vernacular houses were returned to their former glory. One is now a guest house, two are housing the INTACH office and a permanent exhibition on Tranquebar history and architecture. The other two will be devoted to development of crafts.[26]

Given India's colonial past, there is a diverse range of monuments such as forts, churches, cemeteries and so on associated with European presence in India. The urge to tap into this potential as a revenue earner is tempting. However, a note of caution is necessary. Heritage tourism, attached as it is to a systemic commodity production ethos, requires the constant production of heritage commodities and inevitably marginalises possibilities for creative engagements with cross-cultural historical encounters. 'Ultimately it reduces history to "artefactual" history, whereby tourists imagine pasts drawing on what restored or manufactured structures visually convey to them, having carefully excluded the unpleasant aspects of history'.[27] Clearly introspection is required as the country ventures into transnational nominations in its search for a bigger role on the global stage.

Figure 10.4 One of the restored buildings on Goldsmith Street, Tharangambadi.

Source: Courtesy of the author

In the next section I take up the case of Project Mausam, which was first presented at the 38th World Heritage committee meeting at Doha, Qatar in 2014,[28] but has so far remained in the planning stage with the Ministry of Culture and the two institutions under it which have been identified for taking it forward, i.e. Archaeological Survey of India and the Indira Gandhi National Centre for the Arts. The next section examines the potential of Project Mausam to promote a collaborative research agenda that would help establish the study of maritime history beyond national boundaries.[29]

Project Mausam and its Outstanding Universal Value

"Mausam" refers to the season when ships could sail safely using prevailing winds. This distinctive wind system of the Indian Ocean region follows a regular pattern: southwest from May to September and northeast from November to March. The English term *monsoon* came from the Portuguese *monção*, ostensibly from the Arabic *mawsim*.

The etymology of the word signifies the importance of this season to a variety of seafarers. The "discovery" of the etesian or annual winds is attributed to the Greeks, although Indian and Arab sailors are known to have used the monsoon winds much earlier. This regular pattern facilitated the movement of people, goods and ideas across the Indian Ocean, enabling cultural interactions and exchange until steam-powered cargo carriers reduced reliance on sailing ships. These ancient connections were not limited to the coastal regions: they pervaded life in the hinterland and impacted inland communities as well.

Fishing as a subsistence strategy dates from at least 10,000 BCE in coastal areas of the Indian Ocean. A few coastal shell midden, open and cave sites with marine shell deposits dating from after 8,000 years ago have been identified in northern Sumatra, western peninsular Malaysia and Vietnam. At present, many of these sites are found inland (e.g. in Sumatra, on an old shoreline 10–15 kilometres away from the coast), thereby reflecting higher sea levels during the middle Holocene. The spread of Austronesian speakers throughout Southeast Asia has been attributed to their success in boat technology. It has been suggested that around 2000 BCE they already had a boat-building technology based upon 'lashings, protruding pierced lugs, and a hollowed base for the hull with added planks. At this stage, however, they must have adopted their own unique triangular sail and the outrigger construction'.[30]

Fishermen, sailors and merchants travelled the waters of the Indian Ocean, linking the world's earliest civilisations from Africa to East Asia in a complex web of relationships. The commodities exchanged through these networks included a wide array of goods: aromatics, medicines, dyes, spices, grain, wood, textiles, gems, stones and ornaments, metals and plant and animal product. These were transported through voyages and sold at markets or bazaars along the Indian Ocean littoral. Many of the commodities involved had multiple meanings and diverse functions. Spices, for example, were not only used as condiments and for preservation of food, but also played a major role in *materia medica* and ritual practices. Additionally, while trade might have underpinned many of these cross-cultural relationships, the ocean was also a highway for the exchange of religious cultures and specialised technologies. The expansion of Hinduism, Buddhism, Islam and Christianity helped define the boundaries of this Indian Ocean "world", creating networks of religious travel and pilgrimage.[31] The construction of traditional sailing craft involved trade and transportation of wood for planking and coconut coir for stitching from different regions of the Indian Ocean, enabling the transmission and preservation of ancient boat-building technologies.[32]

A distinction may perhaps be made between shipping and maritime trade, though the demarcation is no doubt fine and often the lines get blurred, as sometimes there is evidence for the *nakhuda* (master mariner) transporting his own commodities for trade. The theme may be studied from two perspectives, the two not being mutually exclusive: first, the communities involved in the construction and sailing of watercraft; and second, diverse groups of passengers who travelled across the seas for a variety of reasons. While trade and trading activity has been emphasised in recent writings, the literature on boat-building and sailing communities as also on cultural routes and navigation corridors continues to be inadequate and neglected. Thus, it becomes important to stress the active role played by mobile groups who travelled across the seas, such as seafaring communities, religious clergy, scholars, crafts groups, adventurers and so on.

How was this maritime world conceptualised? A response to this question draws in data from sculptural representations on religious architecture, as also visualisation of the oceans in early inscriptions.³³ Another aspect of the maritime networks relates to the visual topography that provided landmarks to sailors and defined the sailing world in antiquity. This visual topography was characterised by coastal structures, many of them religious in nature that created a distinctive maritime milieu. For example, the 13th-century Konarak temple on the coast of Odisha in India was known as the "Black Pagoda" to European sailors, as opposed to the "White Pagoda", the Jagannath temple in Puri. Similarly, the Buddhist temple at Nagapattinam on the Tamil coast in India, erected for Chinese Buddhists, was a major landmark for ships from the seventh to the 19th centuries when it was demolished by French Jesuits. A theme that remains under-researched relates to the role of the religious shrine in cultural and social integration.

The temples and monasteries were not merely centres of devotion and worship, but were also intermediaries between the householder and the king, as well as institutions for establishing laws and enforcing them on their members. In addition to their role as adjudicators in society, religious shrines were also centres of learning, as well as important locales for recitation of the epics and the *Puranas,* and vital partners in conducting religious festivities. The building of new temples has been seen as a contributing factor in stimulating economic growth, thereby transforming both the geographic and cultural landscapes of the region.³⁴ At the same time there are several instances of a differential tax on commodities required for religious purposes. Eleventh century inscriptions from the temple in Thirumukkudal, on the banks of the Palar river near Kanchipuram in Tamilnadu indicate the existence of a Vedic pathsala attached to the temple, as also a medical centre termed *athura saalai* and arrangements for distribution of medicinal herbs. Inscriptions from the Buddhist site of Kanheri near Mumbai and a temple in Gujarat would suggest that this practice may have earlier beginnings.

In Bhadresvara and at several other coastal sites in Gujarat such as Cambay, Somnath and Patan there are a number of tombstones dated between mid-12th and early 13th centuries. Many of the architectural features at Bhadresvara are also found in a mosque built by early settlers at Junagadh, where an inscription records that the chief merchant and ship-owner Abu'l-Qasim b. Ali al-Idhaji built the mosque in 685 AH (1286–1287 CE). How did this changing religious landscape translate into interactions between communities?

The partnership between the *nakhuda* and local communities is best exemplified by the Somnath-Veraval inscription of 1264 CE from Gujarat, consisting of two slabs of stone inscribed with a Sanskrit epigraph and two months later an Arabic record. Located at Somnath-pattana on the Gujarat coast, the inscription records endowments for the maintenance of a mosque (*dharmasthāna* in Sanskrit) and for providing it other services by *nakhuda* Nūr al-Daula wa-l-Dīn Fīrūz, son of Khoja *nakhuda* Abu Ibrahim of Hurmuja-deśa or Hormuz at the mouth of the Persian Gulf together with the local community leaders (*thākura* in Sanskrit). At that time the king or *raja* of the coastline (*velākūla*) of Hormuja was *amir śri rukana dina,* Sultan Rukn al-Dīn. The Sanskrit version of the inscription provides a detailed description of the land acquired as a part of the endowment. This land was located in the vicinity of the city and at least four local residents are listed as providing the land for the endowment. In addition Nūr al-Dīn acquired and donated the products of an oil mill and two shops or marketplaces (*hattas*). The surplus, if any from these was to be sent to the holy places of Mecca and Medina (*mashāmadinā-dharmasthāna*). This transaction was witnessed by all the *jamāthas* (groups) of Somanath who were also responsible for the upkeep of the property. The groups named are as follows: *Nākhudā-nāvika; gamchikas* or oil men along with their preacher; *chūnakāras* or whitewashers; and also *mushalamānas* or Muslims of the town.[35]

The knowledge and use of the monsoon impacted ancient and historical economies, religion, politics and cultural identity.[36] Centuries of trade, migration, colonialism and modern statecraft transformed these traditional interactions across the Indian Ocean, but present-day national identities and perceptions of the past are deeply interwoven with age-old ties. This intertwining of natural phenomena such as monsoon winds and the ways in which these were harnessed historically to create cultural networks provide building blocks for contemporary societies, as they work towards universal values and trans-border groupings – both of which underwrite UNESCO's 1972 World Heritage Convention. As world history acquires centrality and the focus shifts from national histories to globalisation, the history of the sea is discussed as "connected history" across porous borders, linked through boat-building traditions, community networks and cultural practices.[37]

How would one define the OUV of Project Mausam? Do cultural routes have an intrinsic value of art and heritage? Increasingly researchers have argued that values are not inherent in monuments, sites and routes, but ascribed to them by communities engaging with them.[38]

> The 1994 text of the Global Strategy, in particular, stresses that States Parties should move away from an architectural and monumental conception of cultural heritage to one that is more anthropologically inclined in order to consider heritage in a more holistic and pluri-disciplinary manner. One of the outcomes of this meeting and the accompanying text was the 1994 revision of cultural heritage criterion (i). The reference to "a unique artistic achievement" was removed as it was felt that this phrase favoured aesthetically and architecturally pleasing buildings. From that time onward, cultural heritage criterion (i) has referred to sites that "represent a masterpiece of the human creative genius".[39]

Despite these changes, States Parties continue to identify aesthetic and architectural criteria for inscription of World Heritage Sites. Monumental architecture is seen as an icon of national achievement and heritage is used in nomination dossiers to construct collective national identity. However, heritage can be used both in terms of exclusion and inclusion. For example, the nomination dossier of Vézelay, Church and Hill highlighted links between this property and the Second and Third Crusades, thereby situating France in opposition to the Muslim world. In contrast the inscription of the Old Town of Segovia and Its Aqueduct in Spain represented a type of social structure where different religious faiths peacefully coexisted, including Christians, Muslims and Jews.[40]

The larger issue that this chapter addresses is the extent to which this shift from understanding monuments and archaeological sites as centres of cultural exchanges to projecting transnational cultural routes as heterotopic spaces that underwrite UNESCO's World Heritage Convention of 1972 would impact the preservation and protection of a plural understanding of the past and its legacy in a post-colonial globalised world. The idea of routes as World Heritage Sites was initiated in November 1994 during the UNESCO Meeting at Madrid on Routes as Part of our Cultural Heritage. Academic research in support of this premise would no doubt reinforce the recognition of heritage of cultural routes and the need to include these within academic discourses of the seas.

This shift entails re-establishing the centrality of the sea and viewing it not only as a space that permits movement, but also as a site of intertwined cultural encounters and shared experiences, as articulated through histories of material remains found at archaeological sites. In an earlier publication[41] I have discussed the cultural route that connected the island of Salsette off the west coast of India and the island of Socotra at the mouth of the Red Sea, through an analysis of the rock-cut caves and inscriptions found at the two sites. These were written in the Brahmi script and in Sanskrit and Prakrit languages in the early centuries of the Common Era. It is significant that the cultural route far outlived the time-span of the structures and the inscriptions and continues into use into the present.

Traditional watercraft is employed to this day in a vibrant network between Socotra and the west coast of India. Teakwood for the construction of the *hawārī* boats on Socotra is still imported from India.[42] As documented by an oral history project of the University of Warwick, sailors from Kachchh in Gujarat on their way to East Africa often stop at Socotra to pay homage to goddess Sikotar Mata, offering ship models to her and seeking her blessings. The temple of Sikotar Mata (*padia*) situated on the Sikotar Hill coast is maintained by a Hindu priest or *pujari* and a visit to the temple is an integral part of the beliefs that the seafarers have followed for centuries to protect them from the perils of the sea.[43] Thus several traditions of the past live on. There is an urgent need to include them within the academic discourse not simply for their protection and preservation, but more so for a holistic understanding of sea spaces. Reducing this complexity of interactions merely to a mono-cultural category subsumed under the nomenclature of "trade" runs the risk of undercutting UNESCO's agenda of promoting a plural and multicultural understanding of the past and instead implicating the world body in a narrow promotion of current economic interests of nation states. It is here that academic discussions and collaborative research projects can provide the much-needed corrective to hegemonic undertakings under the World Heritage banner.

It is no coincidence that in 2013 Chinese President Xi Jinping announced the creation of a new Maritime Silk Road during a visit to Indonesia in October 2013. This was followed by a keynote address at the March 2015 Boao Forum for Asia which provided details of China's vision for a new Silk Road Economic Belt and Maritime Silk Road, collectively known as the "Belt and Road". This recourse to history proved useful as Xi envisaged a contemporary Chinese project for the development of a series of major ports on the Eurasian rim between China and the Mediterranean to promote maritime connectivity.[44]

This focus on the Maritime Silk Road connecting the south China coast with Turkey through the sea routes in the period from the 3rd century BCE to the 16th century CE is significant, as also is the involvement of members of UNESCO, ICOMOS and ICCROM in the deliberations to frame the transnational proposal. An Expert Committee Meeting on 30 and 31 May 2017, funded by China and hosted by the University College London, decided to probe five of the themes and three of the sub-regional groupings further. No doubt this proposal is counter to the position of historians with reference to China's participation in the Indian Ocean maritime network. For example, Tansen Sen has argued that the southern coast of China was inhabited by non-Sinitic people. Though the coastal areas were incorporated in the early Chinese polities, the Chinese state had little interest in advancing territorial control into the maritime realm before the 13th century. Information about the Indian Ocean world found in first millennium CE Chinese sources were largely drawn from foreign tribute carriers appearing at the Chinese court.[45] Thus clearly the convening of the Expert Committee meeting to discuss the Maritime Silk Routes in May 2017 was not only about re-creating a narrative to suit the present, but also an attempt by UNESCO's World Heritage Committee and members of its Advisory Bodies to play the role of mediators between academics and States Parties.

Because of the fragmented nature of the World Heritage Conventions, any discussion of maritime heritage tends to focus on material and monumental remains along the coast, such as ports and wharfs, and excludes the navigational knowledge so vital to the charting of routes and creation of communication corridors in the pre-modern world. The traditional system of navigation in the Indian Ocean was based on stellar knowledge, and nautical learning was founded on the accumulated experience of navigators. These skills were communicated orally and learnt during years of apprenticeship. The *Muallim nī pothīs* (captain's manuals) now preserved in the National Museum,

New Delhi, provide fascinating insights into the sailing world of the Indian Ocean and changes over time. One *pothī* bears the date VS 1710, or 1664 CE, which makes its contents the oldest known Indian coastline maps. The inclusion of this navigational knowledge into World Heritage would perhaps only be possible through the Memory of World Register, rather than as a part of transnational heritage.

In the context of both Project Mausam and the Maritime Silk Route Project, the objective is to establish transnational linkages with the focus on the sea. Here I shall restrict myself to the seas that connect India and China. Recent research in the field has brought to light new data from shipwreck sites in the region which provide valuable insights into cross-cultural linkages. These inputs from shipwreck sites are seldom integrated into large discussions on cross-cultural interaction across the Indian Ocean, which remain confined to trade networks of land-based States and empires, as also urban centres and ports or harbour installations. A theme rarely written about relates to religious shrines that dot the coasts of the Bay of Bengal and the South China Sea, as discussed in the next section, and their role in social and cultural integration of communities that travelled across the seas, as also to centres further inland.

Earliest shipwreck site in South Asia and the ritual economy of seafaring

As discussed earlier, holistic understanding of the complex pre-modern maritime network powered by the sailing ship involves utilisation of a range of varied sources such as maritime ethnography, archaeology and anthropology rather than the somewhat restrictive domain of archival data and study of trade and trading activity. In the absence of modern nautical charts and maps, the sailing world was defined either through navigation corridors established by the monsoon winds or by coastal features along the shore.[46] 'The chains of perceptibility created by looking from one vantage point to the next served both to express the relationship of individual localities to one another and to make sense of the wider world'.[47] This maritime milieu created coastal enclaves which dominated communication both across the ocean as well as with the hinterland. An important component of this maritime landscape was coastal architecture and its inter-linkage with travelling groups who moved both across the sea, as well as on routes into the interior. The demarcation of sea spaces may be understood through intellectual traditions of writing, but more importantly through an active engagement with the nature of coastal installations that physically circumscribed the seafaring world and framed the interactions of several groups.

To what extent was the religious shrine or temple a motivating factor in channelling economic activity as also being able to provide anchorage to mobile communities? Patrick Olivelle refers to the social role of the temple as discussed in the *Arthaśāstra* and contrasts it with its absence in the *Dharmaśāstras*.[48] He mentions that references in the text indicate the enormous wealth of temples and their political clout, as evident from the mention of an overseer of temples. In legal disputes relating to temple property, 'we see the temple or the temple god emerging as a legal entity with legal rights that can be defended in a court of law'.[49] Monastic and temple-centred religious institutions formed an important intermediate group between the state and the family. Thus, temples and monasteries were not merely centres of devotion and worship, but were also principal institutions in the period from the 9th to 13th century CE for establishing laws and enforcing them on their members. Nor was the importance of the temple in the cultural life of the community limited to India. Peter Skilling argues that ritual was essential to the functioning of the Thai State: 'Ritual needs influenced trade, since certain ritual paraphernalia – for example the *cāmara*, the whisk fashioned from the tail of the yak – had to be imported over long distances'.[50]

An appropriate example to assess these themes is provided by the earliest shipwreck site so far excavated in South Asia, at Godawaya at the mouth of the Walawe river. On a hillock near the river stands an ancient Buddhist stupa where three inscriptions carved on the rock were found dating to the 1st and 2nd century CE. One of these is the two-line epigraph that mentions that King Gamini Abaya (Gamani Abhaya) identified as King Gajabahu I, who ruled Sri Lanka in the 1st century (113–135 CE) donated the customs duties of the port of Godapavata to the *vihara* (Buddhist monastery) at the site. The small stupa that exists on top of the hill is a major landmark on the coast and can be seen from miles out at sea.[51] Another inscription found at the site records gifts of land by the queen to the stupa. In 2003, an old stone anchor was found at the site, further corroborating its location as a landing site for watercraft.

In 2008 a team of maritime archaeologists from Galle found an isolated wreck, identified through the presence of hundreds of potshards on the sea bed near the Godawaya coast, a small fishing village situated between Ambalantota and Hambantota in southern Sri Lanka. Many of the ceramics scattered on the sea bed show similarities with those found in archaeological excavations at Tissamaharama, 50 km north-east of Godawaya. Other artefacts recovered from the site include saddle querns, glass ingots, copper and wooden fragments. Five examples of rectangular saddle querns of basalt with four legs were recovered from the site; some of these were inscribed with auspicious symbols such as the *nandipada* or taurine symbol, *srivatsa* or auspicious mark and fish. Saddle querns have been found at other Buddhist sites as well, such as at Yatala monastery, about 40 km east of Godawaya and at Ramba, 25 km west of Godawaya. They have also been recorded from several sites in India within the time bracket of 500 and 150 BCE. Archaeological excavations at the fishing village yielded remains of a landing site, as indicated by four stone pillars with an average height of 3.3 m.[52]

According to the Sri Lankan Chronicle, Prince Vijaya and his followers travelling from India landed on this part of the coast in the 5th century BCE. In the 3rd century BCE King Mahanaga founded the city of Mahagama, at present known as Tissamaharama, as the capital of the ancient kingdom of Ruhuna in the southeast of the island. In the 2nd century BCE, a tank was constructed near the site and is known as Tissawewa or Tissa Lake. Between the ancient citadel of Akuragoda and the present urban core of Tissamaharama lies the *mahastupa* and other monastic complexes, while not far from the site are the two coastal landing places at Kirinde and Godawaya.[53]

Archaeological excavations have been conducted since 1992 near the monastic site of Tissamaharama, one of the most revered temples in Sri Lanka. The rigorous analysis of pottery from the site provide valuable insights into the complex relationship between Buddhist religious architecture, the shipwreck site at Godawaya, and also participation in the larger east coast of India network that extended across the Bay of Bengal into Bali and Vietnam.

Archaeological fieldwork concentrated on the site of Akurugoda situated on the eastern bank of Tissawewa has unearthed an ancient citadel surrounded by a rampart, workmen's quarters, and bronze and iron smelting furnaces. A large number of grinding stones and stone tables for herbal medicines indicate the presence of a hospital. The urban layout was preceded by wooden structures directly set into virgin soil around 5th century BCE and was surrounded by an earthen rampart. Though royal power declined by the 3rd century CE, the citadel was never abandoned and the monasteries south of the Tissawewa tank continued to flourish. The oldest imports came from India, such as horses, carnelian and pottery of various fabrics.[54]

The graffito of an ocean-going sailing ship on a potsherd indicates trans-oceanic associations.[55] In addition to the potsherd, the scratched depiction of a ship with a single mast with rigging and twin rudder oars are to be found on a potsherd from Anuradhapura. Similar types of ships are illustrated in the Brahmi inscription of Duvegala. The results from archaeological work at Tissamaharama have provided insights into the participation of coastal centres in Sri Lanka in maritime networks as early as the 5th century BCE, as well as the development of urban centres close to monastic establishments. It is significant that historical memory of landfall by mobile communities survives in the legend of Prince Vijaya as stated in the Sri Lankan Chronicles, which have often been discredited as a historical source.

Nevertheless, it is significant that the legend of Prince Vijaya forms the theme of a painting in Cave 17 at Ajanta, a World Heritage Site in western India. In one of his previous births, Shakyamuni Gautama was born as Simhala, a merchant who led 500 others on a seagoing venture to Tamradvipa or Sri Lanka. They were shipwrecked, but eventually saved from man-eating ogresses by the horse Balaha, who rose majestically into the sky with Simhala on his back. The ogresses, however, followed Simhalaback to his kingdom. Simhala once again rose to the occasion and saved the kingdom from being devoured. Simhala was crowned king and Tamradvipa was renamed Simhaladvipa.[56]

Simhala Vijaya takes on a different character in the Sri Lankan Chronicle, the *Mahavamsa*, which records that Sri Lanka was uninhabited by humans until it was colonised by Vijaya and his followers in the middle of the first millennium BCE (*Mahavamsa*, 7. 1–3). Composed by *bhikkhus* in the 4th to 5th century CE, the *Mahavamsa* was probably compiled from earlier sources. It chronicles the island's past from its colonisation by Prince Vijaya in the 4th or 5th century BCE to the reign of Mahasena (r. 274–301 CE).[57] Archaeological research in the last few decades has established the pre-historic beginnings of settlement on the island long before the landing of Prince Vijaya, though the narrative retained its primacy in the history of Buddhism on the island and linkages with several mobile groups, such as merchants and craftspeople who are also listed as having accompanied Prince Vijaya. The issue is not to seek historical evidence for the Vijaya legend, but to read it as reflection of an ongoing process of integration of different mobile groups by the Buddhist *sangha* (community) on the island. This is a process that links coastal centres in Sri Lanka with those on the east coast of India.

Evidence of the interlinkages between Tissamaharama and the east coast of India is provided by the distribution of a specific ceramic type termed "Rouletted Ware", which was long considered to be a marker for Indo-Roman trade.[58] Painstaking analysis of the pottery from archaeological excavations at Tissamaharama has led to an unbroken chronology and sequence from 5th century BCE to 12th century CE. The results are based on about 4,000 sherds of "Fine Grey Pottery of North Indian Origin". This includes about 2,000 fragments of Rouletted Ware, about 170 sherds of Wheeler Type 10, about 400 sherds of Wheeler Type 18 and about 600 sherds of the Northern Black Polished Ware. Furthermore, around 600 rim fragments of imitation Rouletted Ware were found.[59]

It is now widely accepted that Rouletted Ware first emerged in the 3rd and 2nd centuries BCE and belongs to a group of pottery that has been found at archaeological sites along the east coast of India from Bengal to Sri Lanka from late 3rd and 2nd century BCE to 1st century BCE, though these vessels continued to be kept and used in later time periods also, as pieces of Rouletted Ware have been found with bronze and iron riveting.[60] In addition to clustering along the east coast, Rouletted Ware was also found at sites along the Krishna, Godavari and Kaveri rivers, as well as across the seas both westward and eastward. Its chronology seems to be coterminous with the Mauryan period. Particularly relevant for this chapter are finds at sites in coastal Malaysia, Thailand, Java and Bali, as also Vietnam. Sembiran and Pacung are adjacent sites that extend 250 m inland along a 700 m stretch of the coastal plain of the Batur volcano on the north coast of Bali. Nearly 600 sherds of Rouletted Ware and an equal number of possible coarse Indian wares have been recovered from archaeological excavations in Bali. The arrival of Rouletted Ware at Southeast Asia must have occurred between the 3rd and 2nd centuries BCE, according to the results at Tissamaharama.

The occurrence of Rouletted Ware has been considered as evidence for a direct participation in the Indian Ocean trade in several secondary writings. However, the discovery of Rouletted Ware and its relatives first and foremost indicates only the existence of a site at about the 3rd to 2nd centuries BCE and there is a need to contextualise the finds. The find spots of Rouletted Ware fragments reflect a maritime system along the East Coast and down to Sri Lanka, which appears like the main network along the coast. It shows cross-routings of the peninsula via its main river systems, pathways for the transport of the goods desired either in the West or the East.[61] It is important that finds of Rouletted Ware sherds be placed in context of the site. Was this pottery traded on its own? In which case it would have been reported from market centres and landing places. Since the primary shapes are shallow dishes, would it have been used for serving, eating or cooking? In several cases, Rouletted Ware dishes were inscribed by donors and given as gifts to Buddhist monastic sites.[62]

It is then evident that by the early centuries of the Common Era, maritime travel had acquired complexity and in addition to trade and profits, Buddha *dhamma* or teachings of the Buddha provided it sustainability. The find spots of Rouletted Ware fragments along the east coast of India (Figure 10.5) reflect a maritime system from Sri Lanka extending as far north as Bengal and across the seas to Vietnam and the South China Sea to the east and Berenike on the Red Sea coast to the west. The close association of sites yielding Rouletted Ware sherds along the east coast of India with Buddhist monastic centres is striking. An under-utilised primary source for the spread of cultural knowledge is data from writing on pottery or inscribed potshards: 'Wherever Buddhism travelled, it fostered intimate associations with the written word'.[63] The distribution of inscribed potsherds as early as the 1st century BCE onward and seals and sealings with the ship symbol have been discussed elsewhere to indicate the use of languages such as Sanskrit, Prakrit, Pali, Tamil and Sinhala in the early centuries of the Common Era.[64]

There were nevertheless other communities who participated. Relevant here is the identification of an inscribed small flat rectangular touch stone of 3rd to 4th century CE now kept in the temple Museum of Wat Khlong Thom in south Thailand. The eight letters in Tamil-Brahmi read *perumpataṇkal* meaning "[this is] the [touch] stone of Perumpataṇ". *Perum* means "big" and *pataṇ (pattaṇ)* means "goldsmith". Therefore, Perumpataṇ is a title or the name of the goldsmith who possessed this touchstone.

In secondary writing on the subject, the emphasis has been on trade, which is often termed an "elite" enterprise conducted by ship owning merchants, especially in the second millennium CE.[66] It is seldom appreciated that trade involved a complex range of transactions, with gifts to those in authority and prestige commodities required by powerful groups and residents of cities at one end of the scale, while barter and monetary exchanges were the norm at the local and regional level. The State tapped revenues from trade by taxing the sale of commodities at entry points to the city or in designated markets. Merchants and traders in some cases certainly owned ships and watercraft, but they neither manned nor sailed these. More often, goods and cargoes were entrusted to the captain of the vessel, who was then responsible for their sale and profit. It is also apparent that these diverse groups owed allegiance to a variety of belief systems which continued to evolve,

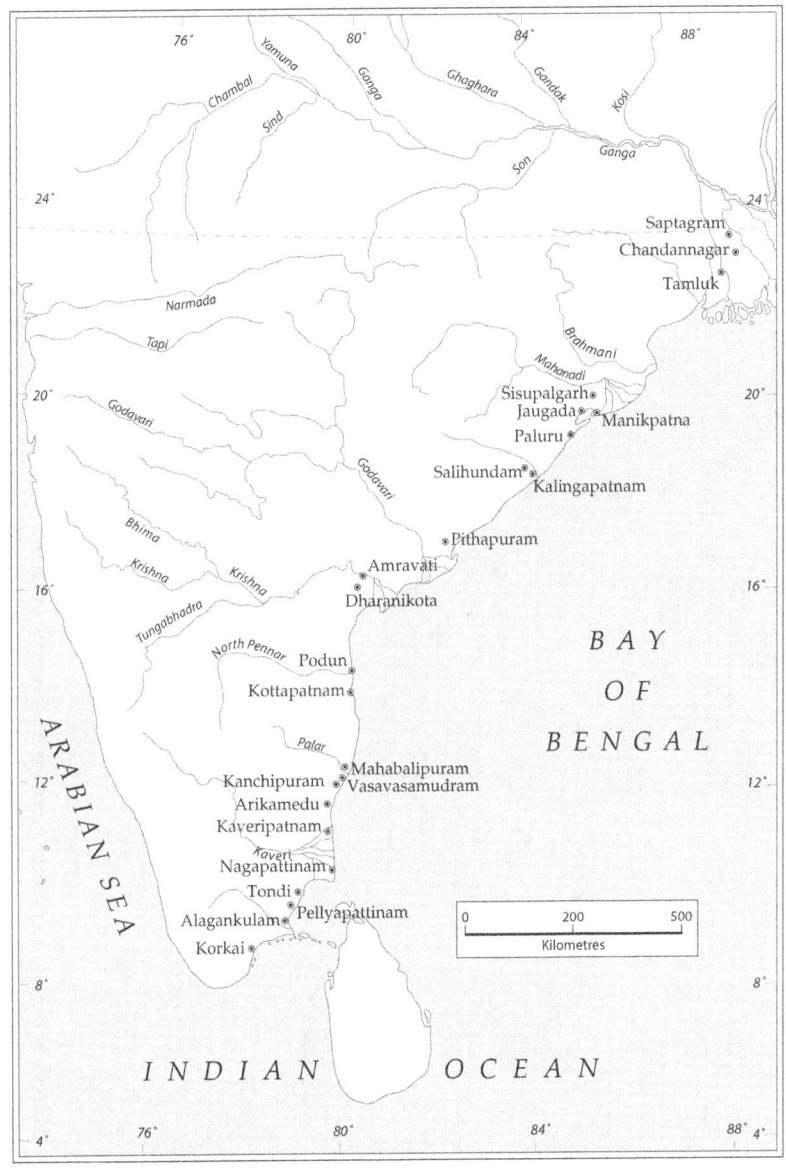

Figure 10.5 Map shon-wing distribution of archaeological sites yielding Rouletted Ware pottery along the east coast of India.

Source: Map drawn by Uma Bhattacharya[65]

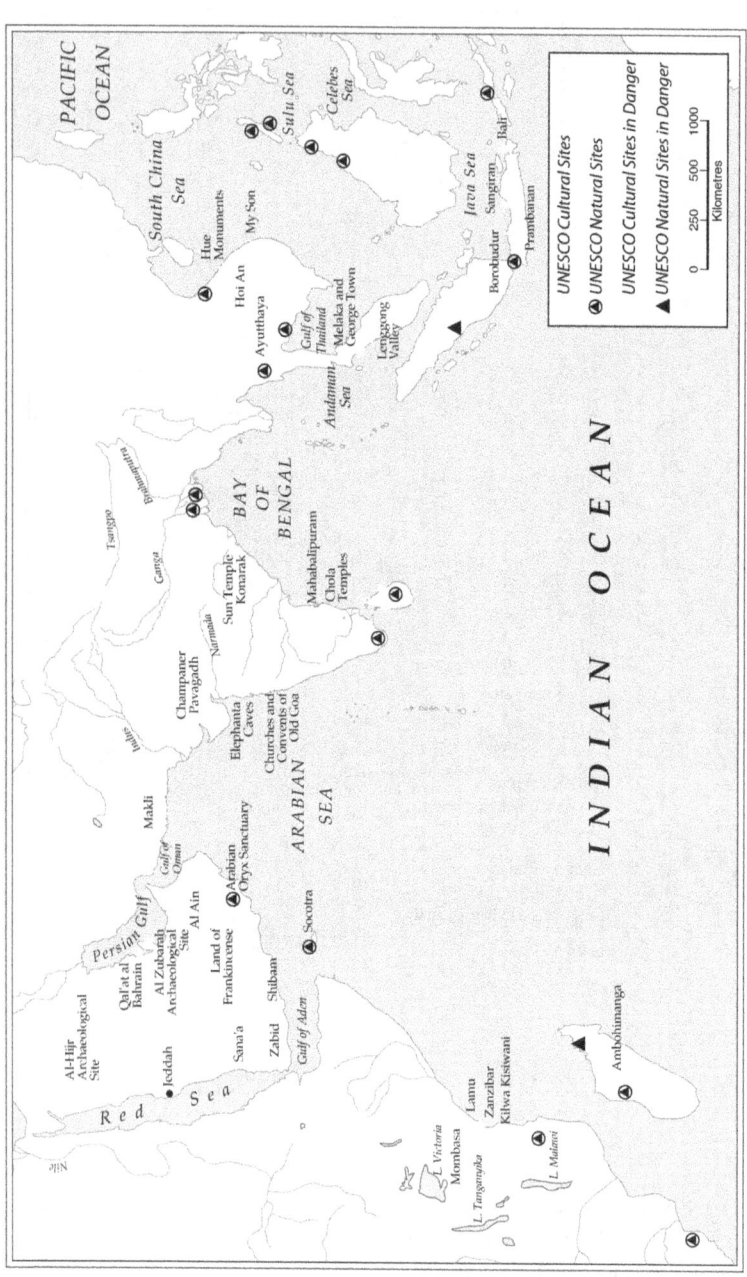

Figure 10.6 Map showing coastal World Heritage Sites across the Indian Ocean.

Source: Map drawn by Uma Bhattacharya

change and co-exist, but which were by no means subsumed or integrated within a homogenising Sanskritic culture, as suggested by Sheldon Pollock.[67] It is the fishing and sailing communities that formed the foundation of maritime activity in the Indian Ocean and provided a continuum to seafaring throughout history, though no doubt their fortunes fluctuated over time. These communities adopted numerous occupations associated with the sea: fishing and harvesting other marine resources, salt making, sailing, trade, shipbuilding and piracy.

A discussion on the role of various communities in movement of culture and ideas must incorporate the role of scholars, scribes, sailors and Buddhist and Hindu pilgrims. It is evident that one must think and work in terms of multiple networks across the Bay of Bengal encompassing diverse social, religious and cultural groups. It is imperative that these networks not be viewed in terms linear flow of goods, ideas and culture from one nation state to other countries of the Indian Ocean world, but rather highlight the equally if not more important role of mobility and cross-cultural movements in the dynamics of cultural exchange. It is within these larger networks that World Heritage Sites as discussed in this book provide stability and identity not only to present nation states, but also maritime communities that face continued threat to their livelihoods and way of living both from increased mechanisation and environmental degradation.

Notes

1 The World Heritage Centre was established in 1992 to act as the secretariat and coordinator within UNESCO for all matters related to the Convention. The Centre organises annual sessions of the World Heritage Committee and provides advice to States Parties in the preparation of site nominations. The Centre, along with the Advisory Bodies, also organises international assistance from the World Heritage Fund and coordinates both the reporting on the condition of sites and the emergency action undertaken when a site is threatened. Lynn Meskell, 'States of Conservation: Protection, Politics, and Pacting Within UNESCO's World Heritage Committee', *Anthropological Quarterly*, 2014, 87(1): 217–244.
2 Kathryn Lafrenz Samuels and Ian Lilley, 'Transnationalism and Heritage Development', in Lynn Meskell (ed.), *Global Heritage: A Reader*, Chichester: Wiley Blackwell, 2015, p. 220.
3 Sophia Labadi, UNESCO, *Cultural Heritage and Outstanding Universal Value: Value-Based Analyses of the World Heritage and Intangible Cultural Heritage Conventions*, Plymouth: AltaMira Press, 2013, p. 77.
4 'Special Expert Meeting of the World Heritage Convention: The Concept of Outstanding Universal Value', http://whc.unesco.org/archive/2005/whc05-29com-inf09Ae.pdf (accessed on 10 May 2014), p. 15.

5 Lynn Meskell (ed.), *Global Heritage – a Reader*, Chichester: Wiley Blackwell, 2015, p. 5.

6 Rasika Muthucumarana, 'Godawaya: An Ancient Port City (2nd century CE) and the Recent Discovery of the Unknown Wooden Wreck', www.archaeology.lk/668 (accessed on 14 October 2017).

7 The Architectural Work of Le Corbusier, an Outstanding Contribution to the Modern Movement, https://whc.unesco.org/en/list/1321 accessed on 28 June 2018.

8 The Architectural Work of Le Corbusier, an Outstanding Contribution to the Modern Movement, https://whc.unesco.org/en/list/1321 accessed on 28 June 2018.

9 Ravi Kalia, 'Modernism, Modernization and Post-Colonial India: A Reflective Essay', *Planning Perspectives*, 2006, 21(2): 133–156.

10 Ravi Kalia, *Chandigarh: The Making of an Indian City*, New Delhi: Oxford University Press, 1987, p. 2.

11 Kalia, *Chandigarh*, pp. 3–4.

12 Kalia, *Chandigarh*, pp. 26–27.

13 Kalia, *Chandigarh*, pp. 26–27.

14 Kalia, *Chandigarh*, pp. 46, 71.

15 Kalia, *Chandigarh*, p. 73.

16 'The Architectural Work of Le Corbusier, An Outstanding Contribution to the Modern Movement', https://whc.unesco.org/en/list/1321 (accessed on 28 June 2018).

17 *Indian Archaeology – a Review 1970–71*, pp. 7–8; *Indian Archaeology – a Review 1985–86*, p. 15.

18 Elizabeth Errington, *The Western Discovery of the Art of Gandhara*, p. 206; The Government Museum and Art Gallery, *Chandigarh, India: A Handy Guide*, Chandigarh: The Government Museum and Art Gallery, 2003, p. 4.

19 S. P. Gupta (ed.), *Kushana Sculptures From Sanghol (1st–2nd Century AD): A Recent Discovery*, New Delhi: National Museum, 1985.

20 V. S. Naipaul, *India: A Million Mutinies*, New York: Picador, 2010, p. 498. I am grateful to Prof. Lynn Meskell for this reference.

21 Helle Jørgensen, *Tranquebar – Whose History? Transnational Cultural Heritage in a Former Danish Trading Colony in South India*, Hyderabad: Orient Blackswan, 2014.

22 T. Subramanian, *Tarangampadi (Tranquebar): Excavation & Conservation Report 2001–2002*, R. Kannan (ed.), Chennai: Department of Archaeology, Government of Tamilnadu, 2003.

23 Himanshu Prabha Ray, 'A Chinese Pagoda at Nagapattinam on the Tamil Coast: Revisiting India's Early Maritime Networks', IIC Occasional Paper, 66, India International Centre, New Delhi, 2015; Himanshu Prabha Ray, 'Trans-Locality and Mobility Across the Bay of Bengal: Nagapattinam in Context', in Shyam Saran (ed.), *Cultural and Civilisational Links between India and Southeast Asia*, New Delhi: Palgrave Macmillan, 2018: 31–50.

24 Raja H. Swamy, 'The Fishing Community and Heritage Tourism in Tarangambadi', *Review of Development and Change*, January–December 2009, XIV(1 & 2): 197–226.

25 www.livemint.com/Sundayapp/QeJX6vwZyYuAGStmh2vZ9K/The-restoration-of-Tranquebars-heritage.html (accessed on 28 April 2017).

26 'Tharangambadi: The Town of the Singing Waves', www.intachpondicherry. org/flipbook/pdf/Tranquebar_Heritage_walk_Map.pdf (accessed on 28 April 2017).

27 Swamy, 'The Fishing Community and Heritage Tourism in Tarangambadi', p. 216.

28 Himanshu Prabha Ray (ed.), *Mausam: Maritime Cultural Landscapes Across the Indian Ocean*, New Delhi: National Monuments Authority and Aryan Books International, 2014.

29 Himanshu Prabha Ray, 'Writings on the Maritime History of Ancient India', in Sabyasachi Bhattacharya (ed.), *Approaches to History: Essays in Indian Historiography*, New Delhi: ICHR and Primus Books, 2011, pp. 27–54.

30 G. Adrian Horridge, 'The Austronesian Conquest of the Sea – Upwind', in Peter Bellwood, J. James Fox and Darrell Tryon (eds.), *The Austronesians: Historical and Comparative Perspectives*, Canberra: Australian National University, 1995, pp. 143–160.

31 Dionisius A. Agius, *Classic Ships of Islam: From Mesopotamia to the Indian Ocean*, London and Boston, MA: E. J. Brill, 2008.

32 Abdul Sheriff, *Dhow Cultures of the Indian Ocean*, London: Zanzibar Indian Ocean Research Institute and Hurst & Company, 2010; Charlotte Minh-Hà L. Pham, *Asian Ship Building Technology*, Bangkok: UNESCO, 2002.

33 Himanshu Prabha Ray, *Maritime Archaeology of the Indian Ocean*, Oxford: Oxford University Press, 2016; Himanshu Prabha Ray, *The Archaeology of Seafaring in Ancient South Asia*, Cambridge: Cambridge University Press, 2003.

34 Cynthia Talbot, *Pre Colonial India in Practice: Society, Region and Identity in Medieval Andhra*, New Delhi: Oxford University Press, 2001, p. 87.

35 Alka Patel, 'Transcending Religion: Socio-Linguistic Evidence From the Somnatha-Veraval Inscription', in Grant Parker and Carla M. Sinopoli (eds), *Ancient India in Its Wider World*, Ann Arbor, MI: University of Michigan, 2006, pp. 143–164.

36 Tom Hoogervorst, *Southeast Asia in the Ancient Indian Ocean World*, Oxford: BAR International Series S2580, 2013; Mark Horton and John Middleton, *The Swahili: The Social Landscape of a Mercantile Society*, Malden, MA: Wiley Blackwell, 2000.

37 Burkhard Schnepel and Edward A. Alpers (eds.), *Connectivity in Motion: Island Hubs in the Indian Ocean World*, Palgrave Series in Indian Ocean World Studies, Basingstoke: Palgrave Macmillan, 2017.

38 Sophia Labadi, *UNESCO, Cultural Heritage, and Outstanding Universal Value*, Lanham, New York and Toronto: AltaMira Press and Plymouth, 2013, p. 13. Labadi suggests that 'Western approaches to understanding and conserving heritage sites, central to the formal discipline of heritage preservation, can further explain the current prevalence of the belief that intrinsic values reside in these places' and that sites have to be preserved in their original form, design, material and workmanship.

39 Labadi, *UNESCO, Cultural Heritage, and Outstanding Universal Value*, p. 60.

40 Labadi, *UNESCO, Cultural Heritage, and Outstanding Universal Value*, p. 69.

41 Himanshu Prabha Ray, 'From Salsette to Socotra: Islands Across the Seas and Implications for Heritage', in Burkhard Schnepel and E. A. Alpers (eds), *"Connectivity in Motion": Island Hubs in the Indian Ocean World*, Basingstoke: Palgrave Macmillan, 2018.

42 Julian Jansen van Rensburg, 'The Hawārī of Socotra, Yemen', *The International Journal of Nautical Archaeology*, 2010, 39(1): 99–109.

43 Nishant and Shivji Fofindi, http://www2.warwick.ac.uk/fac/arts/history/ghcc/eac/oralhistoryproject/resources/nishantandshivji/ (accessed on 5 April 2016).

44 Kwa Chong Guan, 'The Maritime Silk Road: History of an Idea', Nalanda–Sriwijaya Centre Working Paper Series No. 23, ISEAS, Singapore, October 2016, p. 2.

45 Tansen Sen, 'Early China and the Indian Ocean Networks', in Philip de Souza and Pascal Arnaud (eds), *The Sea in History: The Ancient World*, Suffolk: Boydell and Brewer, 2017, pp. 536–547.

46 Himanshu Prabha Ray, 'Crossing the Seas: Connecting Maritime Spaces in Colonial India', in H. P. Ray and E. A. Alpers (eds), *Cross Currents and Community Networks: Encapsulating the History of the Indian Ocean World*, New Delhi: Oxford University Press, 2007, pp. 50–78.

47 Nicholas Purcell and Peregrine Horden, *Corrupting Sea: A Study of Mediterranean History*, Oxford: Wiley Blackwell, 2000, p. 125.

48 Patrick Olivelle, 'The Temple in Sanskrit Legal Literature', in Himanshu Prabha Ray (ed.), *Archaeology and Text: The Temple in South Asia*, New Delhi: Oxford University Press, 2010, pp. 199–210.

49 Olivelle, 'The Temple in Sanskrit Legal Literature', p. 200.

50 Peter Skilling, 'King, Sangha and Brahmans: Ideology, Ritual and Power in Pre-Modern Siam', in Ian Harris (ed.), *Buddhism, Power and Political Order*, p. 183.

51 Rasika Muthucumarana, 'Godawaya: An Ancient Port City (2nd century CE) and the Recent Discovery of the Unknown Wooden Wreck', www.archaeology.lk/668 (accessed on 14 October 2017).

52 R. Muthucumarana, A. S. Gaur, W. M. Chandraratne, M. Manders, B. Ramlingeswara Rao, Ravi Bhushan, V. D. Khedekar and A. M. A. Dayananda, 'An Early Historic Assemblage Offshore of Godawaya, Sri Lanka: Evidence for Early Regional Seafaring in South Asia', *Journal of Marine Archaeology*, 2014, 9(1): 41–58.

53 H.-J. Weisshaar, H. Roth and W. Wijeyapala (eds), *Ancient Ruhuna: Sri Lankan – German Archaeological Project in the Southern Province*, Vol. 1 (Materialien zur Allgemeinen und Vergleichenden Archäologie, Band 58), Mainz: Philipp von Zabern, 2001, pp. 5–39, 5–8.

54 Hans-Joachim Weisshaar, 'Ancient Tissamaharama: The Formation of Urban Structures and Growing Commerce', in Sila Tripathi (ed.), *Maritime Contacts of the Past: Deciphering Connections Amongst Communities*, New Delhi: Delta Bookworld, 2015, pp. 208–228.

55 Weisshaar et al., *Ancient Ruhuna*, p. 16.

56 John C. Holt, *Buddha in the Crown: Avalokiteśvara in the Buddhist Traditions of Sri Lanka*, New York: Oxford University Press, 1991, pp. 49–50.

57 H. Bechert, 'The Beginnings of Buddhist Historiography: Mahavamsa and Political Thinking', in B. L. Smith (ed.), *Religion and Legitimation of Power in Sri Lanka*, Chambersberg: Anima Books, 1978, pp. 1–12.

58 Himanshu Prabha Ray, *Colonial Archaeology in South Asia: The Legacy of Sir Mortimer Wheeler*, New Delhi: Oxford University Press, 2008, pp. 191–194, 199–201.

59 Heidrun Schenk, Tissamaharama Pottery Sequence and the Early Historic Maritime Silk Route Across the Indian Ocean, *Zeitschrift für Archäologie Aussereuropäischer Kulturen, Band 6*, Wiesbaden: Reichert Verlag, 2014, pp. 95–118.

60 Heidrun Schenk, 'The Dating and Historical Value of Rouletted Ware', *Zeitschrift für Archäologie Außereuropäischer Kulturen*, 2006, 1: 123–152.

61 Schenk, 'The Dating and Historical Value of Rouletted Ware', p. 148.

62 Himanshu Prabha Ray, *Archaeology and Buddhism in South Asia*, London and New York: Routledge, 2018, pp. 44–61.

63 Peter Skilling, 'Writing and Representation: Inscribed Objects in the Nalanda Trail Exhibition', in Gauri Parimoo Krishnan (ed.), *Nalanda, Srivijaya and Beyond: Re-Exploring Buddhist Art*, Singapore: Asian Civilization Museum, 2016, pp. 51–100.

64 Himanshu Prabha Ray, 'Multi-Religious Maritime Linkages Across the Bay of Bengal During the First Millennium CE', in Nicolas Revire and Stephen Murphy (eds), *Before Siam: Essays in Art and Archaeology*, Bangkok: River Books and The Siam Society, 2014, pp. 132–151.

65 After Heidurn Schenk, 'The Dating and Historical Value of Rouletted Ware', *Zeitschrift für Archäologie Außereuropäischer Kulturen*, 2006, 1: 123–152.

66 Peter Magee, 'Revisiting Indian Rouletted Ware and the Impact of Indian Ocean Trade in Early Historic South Asia', *Antiquity*, 2010, 84: 1043–1054.

67 Sheldon Pollock, 'The Language of the Gods in the World of Men, Sanskrit, Culture and Power in Pre-Modern India', Berkeley - Los Angeles - London: University of California Press.

References

Agius, Dionisius A., *Classic Ships of Islam: From Mesopotamia to the Indian Ocean*, London and Boston, MA: EJ Brill, 2008.

Bechert, H., 'The Beginnings of Buddhist Historiography: Mahavamsa and Political Thinking', in B. L. Smith (ed.), *Religion and Legitimation of Power in Sri Lanka*, Chambersberg: Anima Books, 1978, pp. 1–12.

Errington, Elizabeth, 'The Western Discovery of the Art of Gandhara and the Finds of Jamalgarhi', Unpublished PhD thesis, School of Oriental and African Studies, London, 1987.

Gupta, S. P. (ed.), *Kushana Sculptures From Sanghol (1st–2nd Century AD): A Recent Discovery*, New Delhi: National Museum, 1985.

Holt, John C., *Buddha in the Crown: Avalokiteśvara in the Buddhist Traditions of Sri Lanka*, New York: Oxford University Press, 1991.

Hoogervorst, Tom, *Southeast Asia in the Ancient Indian Ocean World*, Oxford: BAR International Series S2580, 2013.

Horridge, G. Adrian, 'The Austronesian Conquest of the Sea – Upwind, Peter Bellwood', in J. James Fox and Darrell Tryon (eds), *The Austronesians: Historical and Comparative Perspectives*, Canberra: Australian National University, 1995, pp. 143–160.

Horton, Mark, and John Middleton, *The Swahili: The Social Landscape of a Mercantile Society*, Malden, MA: Wiley Blackwell, 2000.

Jørgensen, Helle, *Tranquebar – Whose History? Transnational Cultural Heritage in a Former Danish Trading Colony in South India*, Hyderabad: Orient Blackswan, 2014.

Kalia, Ravi, *Chandigarh: The Making of an Indian City*, New Delhi: Oxford University Press, 1987.

Kalia, Ravi, 'Modernism, Modernization and Post-Colonial India: A Reflective Essay', *Planning Perspectives*, 2006, 21(2): 133–156.

Kwa, Chong Guan, 'The Maritime Silk Road: History of an Idea', Nalanda – Sriwijaya Centre Working Paper Series No. 23, ISEAS, Singapore, October 2016.

Labadi, Sophia, *UNESCO, Cultural Heritage and Outstanding Universal Value: Value-Based Analyses of the World Heritage and Intangible Cultural Heritage Conventions*, Plymouth: AltaMira Press, 2013.

Magee, Peter, 'Revisiting Indian Rouletted Ware and the Impact of Indian Ocean Trade in Early Historic South Asia', *Antiquity*, 2010, 84: 1043–1054.

Meskell, Lynn, 'States of Conservation: Protection, Politics, and Pacting Within UNESCO's World Heritage Committee', *Anthropological Quarterly*, 2014, 87(1): 217–244.

Muthucumarana, R., A. S. Gaur, W. M. Chandraratne, M. Manders, B. Ramlingeswara Rao, Ravi Bhushan, V. D. Khedekar, and A. M. A. Dayananda, 'An Early Historic Assemblage Offshore of Godawaya, Sri Lanka: Evidence for Early Regional Seafaring in South Asia', *Journal of Marine Archaeology*, 2014, 9(1): 41–58.

Naipaul, V. S., *India: A Million Mutinies Now*, New Delhi: Picador, 2010.

Olivelle, Patrick, 'The Temple in Sanskrit Legal Literature', in Himanshu Prabha Ray (ed.), *Archaeology and Text: The Temple in South Asia*, New Delhi: Oxford University Press, 2010, pp. 199–210.

Patel, Alka, 'Transcending Religion: Socio-Linguistic Evidence From the Somnatha-Veraval Inscription', in Grant Parker and Carla M. Sinopoli (eds), *Ancient India in Its Wider World*, Ann Arbor, MI: University of Michigan, 2006, pp. 143–164.

Pham, Charlotte Minh-Hà L., *Asian Ship Building Technology*, Bangkok: UNESCO, 2002.

Purcell, Nicholas, and Peregrine Horden, *Corrupting Sea: A Study of Mediterranean History*, Oxford: Wiley Blackwell, 2000.

Ray, Himanshu Prabha, *The Archaeology of Seafaring in Ancient South Asia*, Cambridge: Cambridge University Press, 2003.

Ray, Himanshu Prabha, 'Crossing the Seas: Connecting Maritime Spaces in Colonial India', in H. P. Ray and E. A. Alpers (eds), *Cross Currents and*

Community Networks: Encapsulating the History of the Indian Ocean World, New Delhi: Oxford University Press, 2007, pp. 50–78.

Ray, Himanshu Prabha, *Colonial Archaeology in South Asia: The Legacy of Sir Mortimer Wheeler*, New Delhi: Oxford University Press, 2008.

Ray, Himanshu Prabha, 'Writings on the Maritime History of Ancient India', in Sabyasachi Bhattacharya (ed.), *Approaches to History: Essays in Indian Historiography*, New Delhi: ICHR and Primus Books, 2011, pp. 27–54.

Ray, Himanshu Prabha, 'Multi-Religious Maritime Linkages Across the Bay of Bengal During the First Millennium CE', in Nicolas Revire and Stephen Murphy (eds), *Before Siam: Essays in Art and Archaeology*, Bangkok: River Books and The Siam Society, 2014, pp. 132–151.

Ray, Himanshu Prabha, 'A Chinese Pagoda at Nagapattinam on the Tamil Coast: Revisiting India's Early Maritime Networks', IIC Occasional Paper, 66, India International Centre, New Delhi, 2015.

Ray, Himanshu Prabha, *Maritime Archaeology of the Indian Ocean*, Oxford: Oxford University Press, 2016.

Ray, Himanshu Prabha, *Archaeology and Buddhism in South Asia*, London and New York: Routledge, 2018.

Ray, Himanshu Prabha, 'From Salsette to Socotra: Islands Across the Seas and Implications for Heritage', in Burkhard Schnepel and E. A. Alpers (eds), *Connectivity in Motion: Island Hubs in the Indian Ocean World*, Basingstoke: Palgrave Macmillan, 2018.

Ray, Himanshu Prabha, 'Trans-Locality and Mobility Across the Bay of Bengal: Nagapattinam in Context', in Shyam Saran (ed.), *Cultural and Civilisational Links between India and Southeast Asia*, New Delhi: Palgrave Macmillan, 2018: 31–50.

Ray, Himanshu Prabha (ed.), *Mausam: Maritime Cultural Landscapes Across the Indian Ocean*, New Delhi: National Monuments Authority and Aryan Books International, 2014.

Samuels, Kathryn Lafrenz, and Ian Lilley, 'Transnationalism and Heritage Development', in Lynn Meskell (ed.), *Global Heritage: A Reader*, Chichester: Wiley Blackwell, 2015, pp. 217–239.

Schenk, Heidrun, 'The Dating and Historical Value of Rouletted Ware', *Zeitschrift für Archäologie Außereuropäischer Kulturen*, 2006, 1: 123–152.

Schenk, Heidrun, 'Tissamaharama Pottery Sequence and the Early Historic Maritime Silk Route Across the Indian Ocean', in *Zeitschrift für Archäologie Aussereuropäischer Kulturen*, Band 6, Wiesbaden: Reichert Verlag, 2014, pp. 95–118.

Schnepel, Burkhard, and Edward A. Alpers (eds), *Connectivity in Motion: Island Hubs in the Indian Ocean World*, Palgrave Series in Indian Ocean World Studies, Basingstoke: Palgrave Macmillan, 2017.

Sen, Tansen, 'Early China and the Indian Ocean Networks', in Philip de Souza and Pascal Arnaud (eds), *The Sea in History: The Ancient World*, Suffolk: Boydell and Brewer, 2017, pp. 536–547.

Sheriff, Abdul, *Dhow Cultures of the Indian Ocean*, London: Zanzibar Indian Ocean Research Institute and Hurst & Company, 2010.

Skilling, Peter, 'King, Sangha and Brahmans: Ideology, Ritual and Power in Pre-Modern Siam', in Ian Harris (ed.), *Buddhism, Power and Political Order*, London and New York: Routledge, 2007, pp. 182–215.

Skilling, Peter, 'Writing and Representation: Inscribed Objects in the *Nalanda Trail* Exhibition', in Gauri Parimoo Krishnan (ed.), *Nalanda, Srivijaya and Beyond: Re-Exploring Buddhist Art*, Singapore: Asian Civilization Museum, 2016, pp. 51–100.

Subramanian, T., *Tarangampadi (Tranquebar): Excavation & Conservation Report 2001–2002*, Edited by R. Kannan, Chennai: Department of Archaeology, Government of Tamilnadu, 2003.

Swamy, Raja H., 'The Fishing Community and Heritage Tourism in Tarangambadi', *Review of Development and Change*, January–December 2009, XIV(1 & 2): 197–226.

Talbot, Cynthia, *Pre Colonial India in Practice: Society, Region and Identity in Medieval Andhra*, New Delhi: Oxford University Press, 2001.

van Rensburg, Julian Jansen, 'The Hawārī of Socotra, Yemen', *The International Journal of Nautical Archaeology*, 2010, 39(1): 99–109.

Weisshaar, Hans-Joachim, 'Ancient Tissamaharama: The Formation of Urban Structures and Growing Commerce', in Sila Tripathi (ed.), *Maritime Contacts of the Past: Deciphering Connections Amongst Communities*, New Delhi: Delta Bookworld, 2015, pp. 208–228.

Weisshaar, H.-J., H. Roth, and W. Wijeyapala (eds.), *Ancient Ruhuna*, Sri Lankan – German Archaeological Project in the Southern Province, Vol. 1 (= Materialien zur Allgemeinen und Vergleichenden Archäologie, Band 58), Mainz: Philipp von Zabern, 2001, pp. 5–39.

Index